Artificial Intelligence

The Most Updated and Complete Guide

(Take Advantage of the Artificial Intelligence and of Its Arrival on the Market)

Steve Christian

Published By **Oliver Leish**

Steve Christian

Artificial Intelligence: The Most Updated and Complete Guide (Take Advantage of the Artificial Intelligence and of Its Arrival on the Market)

ISBN 978-1-7750979-9-0

No part of this guidebook shall be reproduced in any form without permission in writing from the publisher except in the case of brief quotations embodied in critical articles or reviews.

Legal & Disclaimer

The information contained in this book is not designed to replace or take the place of any form of medicine or professional medical advice. The information in this book has been provided for educational & entertainment purposes only.

The information contained in this book has been compiled from sources deemed reliable, and it is accurate to the best of the Author's knowledge; however, the Author cannot guarantee its accuracy and validity and cannot be held liable for any errors or omissions. Changes are periodically made to this book. You must consult your doctor or get professional medical advice before using any of the suggested remedies, techniques, or information in this book.

Table Of Contents

Chapter 1: What Precisely Is Artificial Intelligence?

The simple answer to "What is artificial intelligence?" is this: it is dependent on the person you are asking.

Anyone with only a little understanding of technology might consider it to be a robot. The argument is Artificial Intelligence is an eerily similar persona that has the ability to behave and act independently.

If you asked the artificial intelligence expert the definition of artificial intelligence (s) they would answer that it's a set of algorithms that produce outcomes without explicitly being told to perform the task. They would have the right answer.

The term itself is straightforward: it's the study of programming machines to do human-like tasks. The simplest definition as well as the oldest, dating back to the 1950s, when computer scientists Marvin Minsky and John McCarthy started to study AI.

Since the past few years, the term AI has grown and is becoming more specific. As an example, Francois Chollet, an AI researcher at Google thinks that AI is closely linked with a machine's capacity to change and adapt to new environments. Additionally, AI is able to apply and generalize skills in the face of uncertainty. "Intelligence is the efficiency with which you acquire new skills for tasks you weren't prepared for," He stated in a podcast for 2020. "Intelligence is not skill itself, it's not what you can do, it's how well and how efficiently you can learn new things."

Why There are Difference Definitions for AI

The definition of AI varies from one the individual for three main reasons. There isn't any consensus in the formal sense that has defined the term AI. Artificial intelligence experts are stuck for phrases when it comes down to the definition of this emerging science discipline.

Then, the work of scientists and authors like Isaac Asimov, Frank Herbert as well as others have expanded the possibilities of what an advanced society could be able to achieve. Artificial Intelligence is clearly presented in works such as I, Robot, that one would think that the technology can be achieved at any time in the near future.

In accordance with Moore's Law this could be the case.

There is also an abundance of misconceptions about the concept of AI since humans underestimate and underestimate what computers are able to accomplish. In particular, computers are able to perform extremely complicated computations. These calculations are typically impossible for human brains to do. The machines, on the other side, are having a difficult to complete even simple task.

In other words, it requires an enormous amount of work to make a computer distinguish between a cat and moth. Humans, on the other hand, find such the task is quite easy and can be accomplished within a few seconds. Then, we are able to define AI using two features:

1. Adaptability

2. Autonomy

It is the ability of a human-made system that is adept at learning from prior events or sets of data. If it is capable of applying the knowledge that it has learned to help in future decision-making situations and situations, then it is said that it's AI.

If a computer system can complete jobs in an environment that is complex with no constant guidance from the user We call this autonomous.

Thus, if a computer system is autonomous and flexible, it is said that it's artificially intelligent. Artificial intelligence is able to replicate human intelligence and then improve upon it.

Chapter 2: The Importance Of Artificial Intelligence

There are many factors that make artificial intelligence vital in our current world.

In the next section, we'll learn about the immense impact of artificial intelligence

1) Automation

2) Return on Investment

3.) Improved the quality of

4.) Acuity of data from different sets

5) Quick and precise analysis and predictions

6) Does risk-taking in place of Humans

7) Available 24x7

Automation

In the traditional workplace, that employs human employees the work process can be steady provided certain rules adhere to. In this case, for instance the Pareto rule states that 80 percent of result are attainable with only 20 percent of the work. A further 20% of effort is able to produce the remaining 20 percent of outcomes.

In a variety of low-pressure jobs Most of the time is spent putting off meeting, rearranging meetings, or waiting for things to occur. While communication is important, it hinders the process of getting it completed. If, for example, you email an employee, you'll need be aware of the amount of time required for them to see that you've sent an email. Also, you'll have wait until they understand and read the message you're trying to convey.

Finally, you'll need to sit and wait for them to compose your reply before sending the email to you. It's a normal process. But, it doesn't necessarily have to be the situation with Artificial Intelligence. There's no wait time. There's no need for bureaucracy. Actually, any job that takes a human one hour or more to complete can be completed in just a minute.

Humans need downtime. They require breaks to head to their homes with the family, as well as time to enjoy recreation things. Humans live a lifestyle independent of their own.

On contrary they are a valuable tool can be utilized. They are able to take on tasks without needing to stop for a moment. When you give them an assignment then that's it. So long as you have an internet connection and electricity The possibilities are limitless.

Everyone can get bored by routine tasks. However, machines in contrast don't have to be bored.' Except for the kind of beings that are sentient AI in science fiction stories They're able to function all day long 7 days a week and 365 all year.

This is the reason, for an organization that is growing it is crucial to automatize. While some might believe robots are taking away their work, it could be said that they are simply fantastic instruments that can help people get rid of boring and time-consuming tasks, to allow them to be focused on what's crucial.

In the following sections, various sectors are already using artificial intelligence to achieve and surpass their goals. The future of AI will be controlled by those who make use of this technology.

Return on Investment

AI is able to provide a favorable return on investment (ROI) through a myriad of methods. Today, all major players in nearly all industries have implemented AI in their plans.

Dominos is the undisputed supreme pizza chain, as an example, makes use of AI to study information and devise an approach to speed up delivery times. The same technology was utilized to determine the amount of amount of time required to deliver orders to customers. it increased the precision of ETA's delivered to customers between 75% to 93 percent.

Even though AI helps businesses predict ROI, as per TechRepublic over half of CEOs in the world believe that AI to be able to make a profit within three to five years.

Improved services

Artificial Intelligence (AI) is a promising technology for improving the quality of digital services and products. Particularly, digital services can be improved and quicker, which can assist companies to meet their goals for customer satisfaction.

As an example, Amazon, the world's biggest marketplace, makes use of AI in order to offer the services it offers to its clients. Its shopping platform is one example that AI can be used to enhance customer service.

If new users are searching for goods on Amazon initially come across a range of items that could, or not be pertinent to the product they're searching for. But, as time passes Amazon's algorithms begin to learn about customer's preferences. They will begin to show increasing and more relevant results for searches.

Following a number of transactions, Amazon's AI can tell you what you may be looking for prior to start browsing.

The customer service department is an additional aspect in which AI is extensively used. A few years ago the idea of calling live customer service representatives was not a sure-fire method.

You must be exceptionally lucky to get in touch with an employee of the website on the internet. Nowadays, it's easy chatbots are able to carry on an entire conversation, and even assist the customer who is in need.

AI Enhances Accuracy of Different datasets

Artificial Intelligence (AI) can take on extremely complicated tasks, and deliver accurate outcomes. Instead of using human-generated data, AI can gather data by itself.

We'll take a look at a vast farm for an example. The farm could produce thousands of data points within the course of a SINGLE day. An experienced farmer with years of experience may only be able to be able to identify a few of these details and act with a guess.

Farmers who have an AI system however is able to keep track of the various inputs. The weather changes as well as soil conditions, diseases and pests, variations in temperature and many other factors are all able to be monitored by these systems. Then, they can be evaluated in order to gain a clearer view of the situation and also to determine the future actions. A farmer can take choices based on accurate data which will lead to an abundance of harvest.

You can buy hybrid seeds that are able to stand up to this year's extreme weather conditions instead of returning to the seeds the previous year in the event of a favorable weather forecast.

The satellite data can be transferred to the tractor unit used by farmers (which is running an AI software) and then the farmer is able to study the shrinkage of leaves on his crop and determine when the crop is taken care of.

Precision farming is a brand new kind of farming that's becoming more popular. Thanks to AI technology, farmers can spot deficiencies in nutrition for plants as well as diseases. The growth of weeds can be honed using sensors strategically placed in specific buffer zones. They can help determine the kind of herbicides that are needed and the they should be used. This can help to avoid the excessive use of herbicides. This can save money as well as protect the environment.

Quick and accurate analysis, as well as predictions

Artificial Intelligence and Big Data can be very effective. The majority of companies collect huge amounts of information. They typically use only a fraction of their data, and save the rest for eventual needs.

They draw whatever conclusions they are able to draw from 20% of information they've for the purpose of applying it for future improvement. All the rest can be stored but cannot be changed since any new information is brought through before stored data can be accessed and analysed.

Large amounts of data can be analysed quickly and accurately through AI that can detect patterns that are present in the data. It can, for instance, help with managing data which involves the following actions:

Data gathering -> Validation of data Storage of data in a secure manner the data. Sorting of the data in order to guarantee timely delivery to the user.

It's much more difficult to accomplish these jobs due to the difficulty of the kinds of data accessible today. The volume of information accessible today could easily overpower the power of a team of employees working on outdated computers. The aid of our AI colleagues in this case. What is the role AI have in this? First and foremost, it is capable of hyper-personalization. It is the process of collecting data and connecting it to a specific profile. As an example, certain characteristics can be assigned to users who are online and have an individual profile, by analyzing the structured information.

Based on this personalized data that is tied to profiles, analysts can gather useful and reliable information that can be used to anticipate future actions of the persona.

Additionally, AI can recognize patterns that appear to be gibberish to humans. AI can detect different patterns and identify whether the signals are expected to be present within the data, or if they appeared in a strange manner, etc. With the help of patterns, AI can identify anomalies as well as things that require immediate intervention. Companies can take swift actions to prevent the occurrence of problems.

Risks themselves in place of Humans

One of the major benefits AI has is this. It is possible to overcome human weaknesses through the development of an AI robot which can take on the risky activities for humans. It could be a journey to Mars and defusing bombs and exploring the depths the oceans, or extracting petroleum and coal and more, the robot can be employed efficiently in all kinds of natural or human-made catastrophe.

Consider this: Have been informed of the fire in the Chernobyl nuclear power plant in Ukraine? In the past, there was no robots controlled by AI capable of helping us reduce the radiation's effects by swiftly bringing the fire to a stop because anyone who came close to the center of the plant would be dead in a matter of minutes. Then, they poured the sand and boron out of helicopters in close proximity.

The Impact of Artificial Intelligence in Today's Society

Artificial Intelligence is altering our way of life and work. The concept and notion that computers are capable of fulfilling tasks typically performed by humans, be it in the field of healthcare, communications or transportation, finance or system development can now be a reality. Computers can accomplish the tasks in a short time.

In the midst of exploring AI and the ways it's currently being utilized in the global context, the issue is what percentage of these tasks are actually possible for computers to complete.

Artificial intelligence is everywhere the world. It's used across a range of devices such as cars, elevators to washing machines, and it is transforming everything we do. Walking around or navigating the city, as an example it is becoming easier due to the advent of smartphones.

The traditional responses and the tasks that were originally performed by human beings are transferred to robots that are able to respond to challenges and work with maximum efficiency. AI is particularly useful when conventional methods are inefficient.

In particular, making teachers' and students schedules could be accomplished efficiently with the aid of computer systems and other scientific methods. As we live our lives in a social world Artificial Intelligence is utilized as a social media. Social media sites that are top of the line including Facebook, Twitter, and Instagram are utilized as marketing tools. Not only to market products, but also for promoting them.

AI algorithms have been designed to make it easier for users to take decisions based on the current information. AI collects information from multiple sources, analyzes the gathered data and makes decisions in response to the. AI systems are unique in their capability to learn and adapt when they take choices.

The latest innovations in transportation have led to significant advances in the field of transportation through AI as well as machine-learning. Semi-autonomous cars, for instance come with functions which inform drivers about imminent traffic delays or road work. Modern technology means that cars (like one like a Tesla) is able to decide on its own driving choices with no human input.

The rise of AI is mostly because AI isn't something that is a distant dream. It is a reality which is present in our lives and has been implemented in a variety of industries, by machines running in the background helping us with our daily tasks such as reading emails or playing music.

Within the field of finance Software is utilized to help make financial investments as well as decisions. The software can look into the personal information of borrowers. The program performs a thorough background investigation of the various characteristics of the person who is borrowing.

It can, for instance, examine a person's credit rating and after determining if they have a high credit score, suggests that they get an acceptable loan. This kind of software makes it simpler investors to take rational decision-making when it comes to financial matters.

AI has helped make banking more secure because the banking industry is using advanced machines that are capable of detecting fraudulent cash and credit cards. In addition, banking services are quicker and easier to use and convenient, they're fast in responding to customers' demands with the advent of banking and ATM apps that are quicker than conventional bank trips.

The field of healthcare has seen a lot of evolution. medical practices has gone far. Health care practices are evolving that range from the old-fashioned usage of herbs for curing illnesses to the integration of biotechnology into health systems.

Health-related information can be accessed via their smartphones without advice from a health specialist, since the industry has quickly integrated this incredible technology in its routines. AI can't replace doctors or nurses, however robots are on the way to changing the way we live.

In the past, AI is also present for customer service in order to ease interactions with clients. Call centers use virtual customer support to respond to questions without having to interact with humans.

The customer service assistant virtual interacts with and assists customers on an automatic basis. AI customizes customer service and interactions by offering suggestions on the most effective way to assist the client.

Disadvantages of Artificial Intelligence

Every bright side comes with one negative aspect. Artificial Intelligence, on other hand, is not without negatives. Let's take a look at a few the disadvantages.

Chapter 3: Various Applications

In this chapter, we'll examine how AI is revolutionizing various major areas in this section.

Artificial Intelligence in Search Engines

Another method to develop artificial intelligence is the application of algorithms with a tendency to adopt a character on their own. AI algorithms are developed to assist in making the right decisions using data already in place. AI blends information from different sources, analyzing the collected information, and then acts upon the data. AI systems possess a distinct capacity to adjust to learning when making choices. This enables AI to be highly versatile. There are several instances where AI is employed.

Artificial Intelligence in the Workforce

Human intelligence harnesses artificial intelligence to change the market for labor. Artificial Intelligence (AI) is currently widely utilized in HR departments throughout many businesses. Recruitment and onboarding AI is used to manage applications and select highly skilled candidates to cut down on time spent during the process of interviewing. HR departments are discovering how efficient it can be to mix AI and human.

Businesses such as LinkedIn make use of the potential of artificial intelligence to aid their users in finding suitable jobs that are highly accurate. It's the reason it's so important that job applicants have a an impressive resume. computers will look up every word in your resume. Even before an employer is able to look over your resume, the computer has likely already looked over your resume, selected you for a shortlist, and then recommended the applicant to the recruiter.

Artificial Intelligence in Marketing

Artificial Intelligence is becoming an increasingly important aspect of the world of digital marketing. The purpose of this technology is to improve customer satisfaction. To determine exactly what drives a client to act huge amounts of information need to be analysed and sort through to make precise customer profiles.

The major social media platforms like Facebook, Twitter, and Instagram are utilized to promote marketing strategies, and not only to market products however, they also help promote the products. Every internet user is equipped with various trackers which can be utilized to modify the information that they have access to. This would all take too much time for the employees of an army and is best accomplished by computers with a high level of power which run artificial intelligence.

With algorithm-based marketing AI can improve sales. The story is true of the time that Target learned that a teenager was expecting before her parents discovered. That's strange, isn't it?

There's an interesting story behind the events that transpired. Through analyzing similar shopping patterns and browsing habits, the retailer was able identify that the girl was in the exact point of gestation using the data-mining technique. The specific technique used to come to the conclusion has been kept in cover, but the process can be re-used.

If, for instance, you purchase a specific calcium supplement, and you decide to purchase some lotions and then cap it off with a bag large enough to hold diapers, it is possible to conclude that either you or a family member are expecting a child. AI could come to the same conclusion by using algorithms.

AI helps to create ads that are tailored for specific people, this is accomplished via machine learning. Consumers' movements on websites are able to be predicted so that they can provide users with content that is personalized. AI software can determine the best way to distribute content through different platforms in order to tailor content distribution. This way an advanced site could look different to each person who visits.

Artificial Intelligence and Education Artificial intelligence has a long-standing tradition of education. It all began in the 1950s, when John McCarthy organized a workshop at Dartmouth College. Since then, the machines are gaining more capabilities that are comparable to humans and are becoming ever more human like and are useful for various fields of learning.

There are today many applications of AI within our education systems. In the first place, AI tools are helping create classrooms with international students across the globe. AI is implemented into applications like Skype and Zoom for excellent video quality as well as track students' issues with connectivity.

It has become simple to schedule gatherings, conferences or even classes online. A good example of this is the utilization of Zoom and has increased drastically due to the effect of COVID-19 around the globe as well as everyday activities.

Additionally, AI helps in the assessment and administration of tests. Since the beginning of time teachers spend an enormous amount of time marking tests. AI is rapidly coming in to replace the time-consuming task. The score of multiple-choice test is becoming increasingly taken over by AI. This is particularly true of schools and universities, as student portals enable students to effortlessly check their marks and keep track of their marks without the need to see their lecturers and professors.

Artificial Intelligence in Gaming

Artificial intelligence also has found its place within the world of games. Gaming has been around since the time that humans first realized the thrill of games. The only thing that has changed about playing games today is the fact that artificial intelligence has provided machines the capability to play against human players. Chess, for instance, can be played using smartphones, which allows a player to compete against an automated AI.

Researchers are currently working for integrating AI for rendering playback. In particular it is the Unreal engine is a powerful 3D Real-Time rendering software. It is able to create movie-like pictures on demand, thereby giving games an authentic feel. The effects of light play, gravity as well as other game-related physics can be recreated within the game with results similar to what you would get.

In addition to being excellent and the games extremely engaging in all ways. 3D technology has assisted by making games appear more real (some even seem surreal) as well as visually attractive.

Artificial Intelligence in Banking

In the world of finance the financial industry, decisions regarding investments and financial transactions are becoming increasingly made by AI software. A few of these financial programs are able to find out more information about the borrower through a an extensive and comprehensive background investigation of the applicant's details.

In this case, for instance for instance, an AI machine can assess the creditworthiness of a prospective borrower before making a recommendation about whether or not to approve or deny a loan request. People with good credit scores are automatically approved for loans with attractive rates of interest. The advanced programs help buyers to consider the rationality of their financial choices and help can save banks a significant amount of dollars on overhead and labour.

AI has helped make the banking sector more secure through using increasingly advanced systems that detect financial crime like the usage of fake credit cards and bills. Thanks to the development of AI-driven ATMs as well as banking apps Banking services are much faster and comfortable, but also simple to satisfy customer needs. These have reduced time spent and shaved long lines in the bank.

Artificial intelligence is without doubt the leader in the market for financial services. Trade signals that travel with the speed of light are being analysed in real-time and control trading actions in a fast-paced manner. The majority of international trade is run by automated systems who operate at a faster rate than human traders would, resulting in decisions that are worth trillions of dollars every year. The world economy is now controlled by AI!

Artificial Intelligence in Soccer and Football

In the world of soccer Artificial intelligence is used to discover insights into the strategy, advertisement training, as well as other aspects of managing a profitable club. Sport is being influenced by a variety of ways due to this. Sensor technology, for example aids to improve the technique of players during training.

Strains and injuries can be reduced by tools that review the data of players and offer details on the type of injury that a person might suffer and even help predict the likelihood of injuries! A few football clubs also hired Video assistant referees (VARs) that look for fouls and offside goals.

Artificial Intelligence in Customer Care

AI is being increasingly utilized in customer service, to aid the work of humans.

The entire call center has been eliminated by means virtual assistants for customer service which answer inquiries with no human involvement. The bot is able to handle needs on its own responding to frequently asked questions and resolving problems that are repetitive without needing to sleep. This can be done throughout the night, and repeat it in the morning.

The AI customer virtual assistant for service communicates with customers under automatized situations. AI enhances the experience of customers and interactions by offering prompt recommendations to ensure solid relations with the customer.

Artificial Intelligence in Medicine

Medical science has advanced a great step forward in the field of technological advances. The healthcare industry makes it easy for individuals to keep track of their health via their smartphone and without the help from a medical professional. This wouldn't have been feasible just a couple of decades back.

Biotechnology is advancing the field of medicine. AI can improve diagnosis and results and is transforming healthcare overall. While it's too early to be able to determine if AI is going to substitute nurses and doctors however, it's a good now time to declare that AI can perfectly enhance the medical experts.

Artificial Intelligence in Law

In the age of machine learning and AI applied across all fields and businesses there is no reason to be surprised that law firms are employing AI methods. Law is a major factor in all aspects of human interactions. Security, business and various other fields depend heavily on a huge collection of laws that covers the law of the land. Partnerships, mergers, as well as contracts are all legal. AI is undergoing training to offer guidance on proper strategies for implementing the law.

Artificial Intelligence in Home Security

The following article examines the ways in which AI affects our security and surveillance the context of how technology has a profound impact on all aspects of our daily lives. The past was when security included a dog having be able to bark very loudly to guard us from thugs and other (mostly canines) threats. They were also famous to break into raucous music competitions that could disturb even the most sleepy people from their peaceful rest. The days of that are long gone (well almost)! Artificial Intelligence could be man's greatest partner in the near term.

It was a gradual process. We first improved our facilities and also secured ourselves with guards and monitor our activities all evening. With the introduction of surveillance cameras with video footage homeowners and companies can keep track of the activities and movements of those living within the region concerned.

Motion detectors are also integrated into surveillance systems for homes that will turn on the LEDs when there is the possibility of an invasion. Modern security systems will silently call for assistance if there are intruders close by before triggering the alarm.

Following a night of rest, you'll get up to see what actually transpired during the time you and your entire family was sleeping. Certain systems are advanced sufficient that they can bypass monotonous, empty sections of footage to ensure that it is possible to view the highlights and not waste too many hours.

If an apartment or house gets attacked by unsuspecting visitors security cameras can give a definitive proof of what really happened. Security systems for homes such as Nest have produced valuable footage which has assisted in the conviction of dangerous offenders and brought them before the law.

AI aids in fighting criminality. Patterns of crime can be detected by the police, and this could assist in investigating any future crimes. In the event of a robbery surveillance footage could be replayed to reveal any suspicious activities that could not have been noticed. It is possible to identify patterns that could be utilized to instruct police officers. A couple motorcycle riders wearing helmets going into stores is typically an indicator of a successful old-fashioned burglary.

Artificial Intelligence in Government Agencies

AI can also play an important role in the government. Human laborers are not able to manage huge amounts of data. Consider, for instance, the military. Data gathered there is vast and nearly always too delicate to leave to human control. AI or machine learning is able to effectively and efficiently process these information while maintaining their classification.

Applications for ID cards and passports are now automated and may be completed online. When your documents are processed then you'll receive an appointment for you to take it at a location of your choice.

Artificial Intelligence in Law Enforcement

In certain states, police officers are equipped with body cameras, which capture all they look at. Voices and even faces are recorded in real-time through the camera. It is very difficult to tamper with evidence since the footage is preserved and utilized for later analysis. Fraud has been previously discovered by witnesses.

Technology advancements have led to the detection of fraud is more and more integrated into industry. Corporation Issuer Risk Assessment detects financial and accounting fraud by using algorithms to monitor the insider trading. Documents are reviewed in order for important information that is known by the term technology-assisted reviewing (TAR).

The witnesses to crime can work through wires which take notes and then transmit the conversations unnoticed to police listening stations. Thanks to AI the person could be convicted based on their voice recordings, fingerprints and any other biometrics.

More and more, we rely on media outlets to inform us about what's happening all over the globe. The media generates content through television as well as computers, and phones. The past was when news and information were transmitted via the word of mouth. As time passed, we depended on letters for information and sometimes even greetings which could take a long time.

The advent of media and images has helped us stay up-to-date on the latest news as well as provide a sharp image of the data to be transmitted. Information can be delivered within a few moments, from live broadcasts to recorded instances.

The advent technology such as AI technology, the way music is performed has also changed. It was only performed at special events in the past. Nowadays, music is viewed as an unifying factor and is readily available. Music lovers can request live to hear the songs they would like to hear from radio stations. Music apps and aggregate sites such as Spotify extensively make use of AI.

Artificial Intelligence in Travel

Let's suppose you'd like to travel for a vacation. The first step most people take is search for the destination online. If the location meets what you expect, then your next task is to book a trip. Reservations can be made either by calling the location or through online reservations.

In the present the Artificial Intelligence (AI) chat-bots can be employed to aid customers in answering their inquiries without the need for humans to intervene. AI is used in the hotel industry by using computers that ease the burden of employees.

The drying and washing of clothing in hotels are performed by automated machines who perform these jobs quickly and effectively. Customers are receiving personalized support through robotics which are interacting with in order to obtain details. "Connie," which was adopted by Hilton hotel, Hilton hotel is an excellent instance of an AI robotic assistant in the industry.

Artificial Intelligence in Reproductive Health

The process of infertility can be a stressful and stressful process for couples trying to become pregnant.

When women were unable to conceive then she turned to the traditional herbal remedies to find a solution. The advancement of technology has led to IVF (also known as IVF, or in vitro fertilization. The process involves transferring viable embryos of a patient suffering from infertility out of the body, for instance at a laboratory into the host. Following fertilization, the embryo is then transferred to the infected uterus. It is possible because of an advanced technology that permits the egg to remain in place until it is reintroduced back in the human body.

In the present, a lot of studies are being carried out in order to enhance this method. The fertility cells that are used for IVF is able to be selected precisely and then used to fertilize using AI.

Artificial Intelligence in Transport

AI as well as machine-learning are driving significant technological advancements in transportation. Semi-autonomous cars, for instance include technologies that warn drivers of imminent traffic or if there's construction work on the road ahead. Thanks to the latest technology, vehicles can take and handle every decision regarding steering without requirement for human intervention.

Artificial Intelligence in Agriculture

First time in agricultural history, AI is making it possible to conduct research based on evidence. Artificial intelligence can now be employed to anticipate and determine when the crops are going to be harvested. This includes when it will take for the ripening process of carrots and the time they will be harvested fully, resulting in a higher efficiency of farming.

The evidence-based approach isn't just regarding crop cultivation, as it is also about monitoring and analysis of soil by using agricultural robots as well as predictive analysis. Regarding the growth of agriculture monitoring soils employs the latest methods and techniques for collecting data on the field in order to ensure the health of crops for efficient and sustainable agriculture.

Agriculture is currently benefiting from AI specifically in greenhouses. It is used in the areas of automation models, optimization as well as simulations. In greenhouses, this concept is utilized to estimate the time it takes for various plants to mature in order to harvest them efficiently in order to increase the value of the market.

Two characteristics of AI have become more important in the field of agriculture:

1. Automation

2. Simulation

Machine automation is the collection of concepts and techniques designed to remove the necessity for human intervention during the process of manufacturing. It is used in farms and involves the utilization of numerous control systems and machinery within farms including furnaces, boilers phone networks, vehicles, aircrafts, robots as well as other types of applications.

Simulation mimics actual-world events through the use of models that represent the modifications to the system and process's behaviour. Simulation allows for prediction of how model will change as time passes. The process makes use of the idea of modeling scientifically of natural systems, to get crucial information about what is likely to take place in the near future.

The application of artificial intelligence in agriculture can bring many advantages. It first of all reduces the time and energy required. When farms were in the medieval period, they required a lot of workers with harvesting as well as keeping the bookkeeping for the farms. Many processes were completed prior to the harvest being completed and wasting lots of time as well as interrupting the process of marketing the harvest.

Thanks to the advent AI, work previously requiring hundreds of people can be done using a single machine, resulting in more efficiency and with reduced time. Speedier operations can help to maintain the time frame and consequently, guarantee that products are delivered to the marketplace in good shape.

The second reason is that the application of AI reduces the price of production. There are hundreds of workers who require higher pay for different processes like trimming, weeding picking, transporting, and harvesting. With the help of AI technology, the cost of labor is lessened and harvesting is more efficient and less expensive. If it is used efficiently and in conjunction properly planned it is not a wasted production and is less expensive.

The amount and quality of the agriculture yields have increased. Farmers have now a full overview of their production. AI has also transformed the way farmers work by giving them the benefits of optimizing. . This is possible through the identification of areas that need more care in areas such as the management of pesticides, irrigation and fertilization.

The population grows, and also increases the need for food. In an effort to keep up with the increasing need for agricultural goods, we'll need a larger percent increase in the productivity of our agriculture. Utilizing these advanced technology and artificial intelligence can assist us in reaching our objective to meet the increasing need of food. Through these advancements we can enjoy a huge supply of food, and prevent a crisis in food supply.

Chapter 4: What Is Machine Learning?

Machine learning is the term used to describe computer programs that are able to improve itself automatically through practice as well as the use of data. Machine Learning is a branch of computer science as well as artificial intelligence, which focuses on the application of algorithms and data that simulate the way humans learn, and to increase precision as time passes.

Machine Learning is an crucial aspect of contemporary research and industrial. It employs neural networks and models to assist computers in improving their efficiency over time.

What is a Machine, and is it Possible for Any Machine to Learn?

If we refer to "machine" we are not discussing exotic robots similar to when we're talking about robots inside the classroom when we employ the expression "machine learning". In this case the term "machine" refers to an application that runs on a computer. analysis software.

However, how can machines be taught from the beginning?

A computer simply "learns" by searching for patterns among large volumes of information. If it finds patterns, it alters its algorithms to accurately reflect its findings "truth" it finds. The more data you give the machine, the more intelligent it gets. Once the machine has identified sufficient patterns it begins making its own predictions.

A Short History of Machine Learning.

Arthur Samuel was a pioneer in the field of artificial intelligence as well as games on computers who worked for IBM within the United States. He coined in 1959 the phrase "machine learning. The computer scientist who was enthralled created a program to play games of checkers during the 1950s. However, it was necessary to begin the alpha-beta cutting process since the program on computer had only an insignificant amounts of memory.

He included a scoring system within his concept through the positioning of pieces across the table. The scoring feature aims to determine each team's chance of winning. The program selected its words by employing a method known as Minimax that later evolved into the Minimax algorithm.

Actually, he performed an extensive model of "what-if" analyses, where computers decide to take a step to get the most optimal outcome.

Arthur Samuel also developed additional techniques, which he termed"rote learning. This was eventually replaced with machine learning. The program was able to keep track of all positions it observed and to integrate them with reward function's value. Through recording, storing, and then applying its knowledge to how to play the program was able to learn.

Frank Rosenblatt created the perceptron in 1957, at the Cornell Aeronautical Laboratory by combining the concept of Donald Hebb's brain interaction alongside Arthur Samuel's Machine-learning. In the beginning it was thought that the Perceptron was designed as a tool, and rather than a programmer. It was programmed into the specific machine designed specifically to recognize images, dubbed"Mark 1's Perceptron. Mark 1 was the first to be developed successfully as a neuro-computer.

However, this title was only for a short time as the Mark 1 began to develop issues. It wasn't able to identify the various types of patterns that appear in the visual world, which is the most important feature that allows the machine to become more intelligent. This caused frustration for the programmers, and research on neural networks came to an end because of it.

A different algorithm, The Nearest Neighbor Algorithm which was developed in the year 1967 and utilized for routing mapping. It was among the earliest algorithms to determine the most effective route that was used to resolve the problems of every salesperson. Imagine the early versions of Google Maps on steroids, and you'll have a clear idea of the NNA could be.

In the 60s, the development and utilization of multilayers set the stage the dawn of a new phase in research on neural networks. The discovery was that perceptrons having at least two layers could provide superior processing capacity than a perceptron with a single layer. Multiple layers led to the concept of backpropagation, as well as feed-forward neural network.

In the 70s, backpropagation was developed in the 1970s. The phrase "backpropagation means" refers to the processing of an error in the output, and later returned to the layers of a network for learning for purposes. This allows networks to adjust to changing circumstances through the adjustment of the hidden layers of neurons. In the present deep neural networks are being trained with backpropagation.

The layers that are hidden within the hidden layers of an Artificial Neural Network allow it to be able to handle greater complexity than the older perceptron. They're the primary device to aid in machine learning. The hidden layers of their algorithms are extremely effective in identifying patterns that are difficult for human programmers to recognize. Human programmer will not be able to discern patterns which means they won't be able teach the computer to spot these patterns.

Artificial Intelligence and Machine Learning have taken different paths. Up until the end of the 1970s and the early 1980s at which point it was put on hold, Machine Learning had always been used to train system to train AI. AI research was focused on logic-based approaches instead of algorithms. Additionally, researchers within the areas of computers as well as artificial intelligence retreated from the study of neural networks. This has led to that the Artificial Intelligence and Machine Learning areas to split.

Over the course of a decade during that time, over the course of a decade, Machine Learning Industry reorganized into an entirely different business and struggled to keep its feet on the ground. The focus of the company changed from training in artificial intelligence to dealing with real-world issues with manufacturing services. The focus shifted to strategies and techniques that can be utilized to study probability theory as well as statistics using AI-inherited techniques.

As time passes, Machine Learning researchers have succeeded in developing the ability to recognize facial and voice. They have become an integral component of every day technology like image tagging on Facebook Face unlocking software on mobile phones, and voice recognition assistants for customer service.

What's the point for machine-learning?

Machine learning has been the main driver of technological advances that include:

Chatbots

Auto-driving cars

Tools for analytics

New robot algorithms

Unsupervised and supervised learning

Below is a comprehensive list of the 7 most popular applications of machine learning within the world of business:

Product suggestions

Analyzing sales data

Fraud detection

Dynamic pricing that is determined by the need or demand

Natural processing of language

Systems for learning management

Mobile-specific personalization for better the experience of users

What's the main difference between deep and machine learning?

The differences between deep and machine learning can be summarized in this chart below.

DEEP LEARNING MACHINE LEARNING

When it comes to Deep Learning, algorithms are organized in layers, creating the neural layer which is that can make choices on its own. Machine learning employs algorithms to study and learn from data using the learned knowledge to make an informed decision.

The majority of the information is required for training. Training doesn't require an excessive amount of information.

It is more time-consuming to develop deep learning. It only takes a short time to learn.

- It's developed on an CPU to properly train. It requires the use of a the CPU in order to learn.

Output can come as audio or text. Digital output can be used in format to allow scoring or classifying.

It can be tuned by a variety of ways. There is a limited tuning option.

Neural Networks is a subfield of Machine Learning and is a included in Deep Learning models. They are designed to study information and draw conclusions regularly similar to humans. Machine learning, which we all have a good understanding of, is a system of algorithms which analyze data, develops knowledge from information, and then make decision based upon the learned information.

How Search Engines Use Machine Learning

To fully comprehend the relation between these two concepts We must first understand the definitions for each. The simplest definition of machine learning refers to the capacity of machines to operate in the same way as humans do. Search engines are programs which help users find relevant details via the Internet through the keywords. Google is one such search engine, and so are Duckduckgo as well as Bing.

Just like a fish goes to the ocean Machine learning has now taken into the realm of search engines. How users search for and use information changes constantly. The majority of people prefer instant results that match their needs right in the current moment of their searching. For instance an academic task. Students go to for answers on the Internet for help because they need to finish their work completed quickly so they can get ahead of other members of their family.

Search engines that deliver results that are contextual would be the perfect solution to search for a student as who is in this example. What is the way that the search engine determine which results to provide? A swarm of algorithms have examined the patterns of search results in billions of searches (data) and analysed the most effective outcomes (a mountains of data).

This is why the process of optimizing search engines (SEO) is now becoming important for machine learning. Web sites and content producers strive to make their posts (and different types of information) to get the attention of the search engines while displaying their results.

It has been in use for a while, but it was only recently changed by the Google and other search engines. In particular, Google has adapted to machines learning technologies that as time passes and responds to questions. It is achieved by implementing various algorithm changes through time. Every person who is a self-respecting SEO analyst has heard of updates to the Panda, Penguin, Hummingbird, Medic, Bert and Core changes.

Results from search are more efficient and more reliable than before. It is a popular choice for the majority of users of the internet and with Google getting the top position with a high percent of users per day.

The quality of websites is determined by their position on the search engines. Web pages can be optimised to be optimized for search terms. What is the best way to make an effective search engine get beyond the simple search for keywords?

This information hasn't always been available to us. In the past pertinent information was only available within books or documents found within libraries. Technology advanced and search engines such as Alta Vista came along, they were so ineffective that they influenced contemporary search engines.

Then, Google and other sophisticated algorithms came. These are some of the most effective devices that have proved invaluable and can often ease having to flip through pages looking for the information we need.

Machine learning allows for personal search results much easier to find and makes search distinct to the person. It is the main issue to consider the way they accomplish this. It's almost impossible to discern the difference between this and the magic. Consider the scenario of two people looking for the word bowl. These predictions are personalized to the individual's needs by the information which has been fed to their computer throughout the course of the duration of. If the person is in the construction sector The search engine will be able to determine exactly which options to recommend. If the person is a farmer, they'll find results tailored to their needs that are suitable for farmers. Google will try to figure out the meaning of what you're trying to convey and provides you the data you're searching for.

Machine learning can be efficient in identifying the products and contents. Machine learning can classify products and the content available to users, by analyzing the information which is already available. A good example is when You go to the grocery store. In most cases, you browse through the aisles that you're interested. The reason is that you have an idea of where you can find the particular item.

Similar principles are applied online based upon your previous searches. By using cookies, as well as other mechanisms for tracking Users often receive suggestions and ads based on the previous search results. This makes it easier to search as well as product details.

The analysis of the head and tail is employed by search engines to rectify wrong orders, misspelled words as well as improve the user experience. It is possible because of the use of advanced machine learning. Did you type something online, committed mistakes, then immediately rectified? Random computer programs aren't capable to do this. However, it does reduce the frustration of searching for something on the internet and then not being able discover it.

Analytical tools were used to determine what users are seeking and the web-based stores they are frequenting. They assist in determining the kind of person they're dealing with. Search engines focus on how to lead users to quality content. The engine makes use of keywords to deliver appropriate results in the least possible amount of time.

In the majority of searches the search engines provide many results. However, users usually select the top three. The search results are individualized to the needs of library patrons.

Another place where machine learning plays a significant role is meeting the requirements of clients who are online. Happiness of customers is the main aim of all businesses. Machine learning can provide insights to customers by supplying FAQ pages which provide answers to commonly requested questions.

It is a great idea for a store that sells online. This can save cash (by getting rid of the requirement for many assistance with hand holding) as well as improving customers' experience. The FAQ pages give the answers to queries such as what time frame it will take to fulfill a particular order, the material of the clothing, and so on.

It's been demonstrated how machine-learning is able to perform human-like functions. Let's look at how search engines identify the bad actor pattern. Machine learning can be utilized to detect spammy content that people consider insignificant. This kind of content is spotted by the machines. In the case of an item of data is repeated, the person could decide to stop the information since they do not interact with it.

Another area of importance in which machines learning is required is the analysis of photos. This is because of the amount of pictures that are uploaded on the Internet each day. In order to comprehend the images ML employs image search. By analyzing the content, it can identify the content that may be deemed as offensive. Google provides a function that lets users browse for images and engage using the search results.

If you search the internet to find results that don't include keywords, Google can to find alternatives. As an example, if you search on the internet for "action", "caution" could be found. It highlights synonyms so that it can recognize the words. Machine learning allows this to be done.

If you frequent websites for shopping online or fitness programs, you'll find ads that advertise the items you're seeking. The effectiveness of ads appear will be contingent on the kind of information you're looking through the Internet. The amount of relevance and the quality of ads has been affected through machine-learning.

Chapter 5: Deep Learning

The words "learning" and "intelligence" are interconnected. The idea is that anything capable of learning is smart. Any thing that's considered to be intelligent must be able to comprehend and implement its knowledge and other lessons. The goal of learning is to improve your behaviour in the near future, based on your past experience.

Then, what exactly do you know about deep learning? It is an artificial intelligence type that mimics human behavior to learn. It allows computers to gain knowledge from different experiences and view the world in as to ensure that the Deep Learning approach may be employed to address a vast variety of issues in a wide range of sectors. The application of deep learning has had a profound influence on many areas over time. It has caused significant disruptions to every field that it has been applied to because it is the creation, representation and utilization of basic knowledge that humans can easily access.

Then, what's the importance of deep learning? The ability of deep learning to analyse data is particularly effective for processing data with no end. Facebook For instance, recognizes billions of users with no humans to intervene. They can also recognize handwriting when they've had enough information to make their own assumptions and determine things that are different from each other and also what isn't.

Deep learning dates through the 1950s. Alan Turing foretold the existence of supercomputers that had human-like capabilities. In the mid-1960s, we saw the breakthrough in Deep Learning when Alexei developed neural networks. These have been further refined and expanded through time. What is the ancestor of deep-learning? In the year 2006, Hinton coined the phrase "deep learning".

The human brain is believed as the most powerful source of ideas to develop AI. The brain's initial stages of development and put it into deep learning. Deep learning takes its inspiration of the brain a similar manner that our brain's process of training does.

Deep learning's process of model creation is comparable to the initial learning phase of human beings. Because the human brain is among the most powerful machines, computers are prone to mimic the ways that the brain operates in order to build intelligent machines that do not just complete the work to us, but do the job better and faster. Deep learning has helped bridge the differences between human and computer capabilities. In the end, day, when developing computer systems, the objective is to design them in a way that makes them more like humans.

Our machines should be able to accomplish tasks for us, ranging including driving our cars, and providing customer service.

As a result it is clear that deep learning has a dominant position as a type of technology that's constantly changing and will be around for many years in the form of a tool to improve the capabilities of humans. Transcendence is a great film that demonstrates the way in which AI integrates with humankind. Highly suggested.

Chapter 6: Neural Networks

Neural networks are an array of techniques that mimic the functions of human brains to identify patterns. They process sensory information using an artificial process by labeling or grouping raw data. They recognize patterns that are digital, and they're contained within vectors, into which any real-world data, such as text, images, sound or time-series, have to be converted.

Neural networks assist us in clustering and categorize. They can be thought of as a classifying and clustering layer that sits on top of information you manage and store. They assist in clustering unlabeled data by analyzing the similarities of sample inputs and categorize data with an appropriately labeled data set to work on. (Neural networks are also able to identify features and pass them through to different algorithms to help cluster and classify data which is why you should consider the deep neural network as part of broader machine learning algorithms which include the reinforcement learning, classification and regression techniques.)

Artificial neural networks can be found in a variety of uses in various sectors. There are three general kinds of tasks that the neural networks are able to perform.

Data Processing

Neural networks can be used for making it easier to sort, filter, compress data and clustering it in addition to blind signals separation.

Classification

Neural networks can categorize the data in order to identify patterns which are almost impossible human eyes can detect. They can also detect new the data and take sequential actions.

Approximation of Mathematical Functions

Artificial neural networks do time-series analysis, and therefore make predictions upon past events. They can also conduct regression analysis to find which of the lines is most effective in fitting (which will be the most effective in the process of making decisions).

Neural networks have a lot of value when it comes to Artificial Intelligence and Machine Learning. They've played an crucial role in today's AI technology and will likely play a role for the foreseeable future.

Chapter 7: Robotics

A brief history of Robotics

The first explanation for an anthropomorphic robot figure can be found within the Lie Zi text. The mechanical figure was presented to Zhou's king Zhou by an engineer with a mechanical background called Yan Shi, according to the text. The date was 1023 B.C. Some time later, about 420 BC there was a builder who constructed a wooden bird with wings that could fly. It had an engine powered by steam.

The first century was when Heron of Alexandria included more than 100 automated machines in his work Pneumatica. The automata comprised an engine for fire and a machine for coins. Steam could be used for powering the engines.

Al-Jazari invented a robot that was humanoid 1206 that could be controlled by the help of a band. He also developed automated machines such as peacocks that can move, and hand-washing machines.

Leonardo Da Vinci was the one to come up with a concept for the idea of a mechanical knight. Da Vinci conceived intricate designs for a human-machine robot. The 1560s were when he developed a mechanical monk during the 1560s. Monks could walk with machine feet, which were covered by a cassock that allowed the monk to emulate human walk. Eyes could be moved realistically while the lips swung between up and down.

Then let's go to 1898. Nikola Tesla exhibited the first instance of teleautomation, when the radio was used to steer a vessel. Then, a few years later the Czech writer Karel Capek wrote a piece named Rossum's Universal Robot. The term robot is mentioned in the very first place in human history. In Czech the word robot means labour that is forced.

The robotics depicted in the film were composed from chemical batter. They performed significantly better than human beings in manual work surpassing their human counterparts in the process. However they had a rogue attitude and later carried out the murder of their human masters.

The first robot with a humanoid body was developed in the year 1930 by Westinghouse Electric Corporation, an older American industrial company. The robot was displayed during the 1939 and 40 World Fairs.

The first robot commercially available was created by George Devol and Joseph Engelberger in the year 1956. They later founded an enterprise called unimation and made the first industrial robotics. In the year 1961, General Motors installed its first Unimate industrial robot to support the assembly of its vehicles.

The Unimate progeny is now in charge of the global automobile and manufacturing assembly facilities. They're capable of dangerous tasks, repetitive work as well as precision machining, at an unbeatable level of accuracy.

WABOT-1 was designed by a small group comprising Waseda University academics over a five-year period from the year 1967 through 1972. The WABOT-1 was a humanoid robotic that had some level of intelligence. It was able to walk on its legs, and hold objects using its hands. The sensors are inside its hands, which allowed it to calculate its weight. The eyes and ears of WABOT-1 also functioned. WABOT-1 was able to talk with its eyes and ears in Japanese.

It was the KUKA Robot Group, a German company, released an industrial robot in the year 1973. The name was "Famulus," which had six Axes, which were electromechanically operated. In the year following and it was the IRB 6 was launched by ASEA the Swiss Research and Development firm. The IRB-6 was the only one in its integrated microcomputer.

There's been a flurry of activity within the field of robotics since the early 1980s until today. Honda developed the P3 humanoid robotic in the late 1980s. The robot could walk, shake hands and wave its arms at people effectively. P3 transformed into Asimo that became well-known due to its football-kicking capabilities.

The robotics industry had advanced to a great extent before the decade of the 1990s and 2000s came along due to the research conducted into three technology areas:

Artificial Intelligence

Actuators

Sensor technology

Robots come in various dimensions and shapes, as well as they can be found in all "walks" of life, should you wish to. They perform different tasks and operate in various environments. In spite of their distinct functions however, they share three fundamental features in common:

1. In all robots the form functions. Each robot comes with mechanical components that help in the execution of their task. As an example, Boston Dynamics makes dog-like robots (appropriately known as Spot) that have four legs, which allow the robots to move and behave as dogs. However those that get rid of radioactive waste are equipped with caterpillar tracks, which enable them to move around obstacles. Each robot is equipped with frames that have been adjusted to the various tasks they perform.

2. They are made of electronic components. They manage the machinery and assist in distributing energy to the parts that move. The majority of robots today use electrical circuits either to begin the combustion

process when they're petrol-powered or for consuming charge from batteries in the event that they're battery-powered. Sensors that comprise other electronic components enable robots to monitor the changes that occur in several factors, like temperature, location as well as sound. They also measure the energy stored within the battery.

3. Every robot that exists currently are controlled by codes. Whatever the design an item is it must have the code to tell the robot what it should do. It would not be able to work without this program.

Robots are programmed with three primary types of programming methods. The three main types of coding include:

1. Artificial Intelligence

2. Programming remote control

3. Hybrid programming.

1. Artificial Intelligence (AI)

It is possible for a robot to operate using artificial intelligence if there is programming embedded in the memory chip and chips, and performs its tasks with no human input. Spot is the Boston Dynamics dog, can look around and determine which action to take by using internal code. It is able to attempt to finish tasks without involvement from humans when it has a clear focus upon it.

2. Programming remote control

The programming for remote control is like the programming employed for RC automobiles. All artificial intelligence functions are encoded in the control unit of remote. A signal receiver is integrated inside the robot, and it is able to receive commands via electromagnetic waves. The receiver is then able to interpret the signal and then executes exactly what the human controller wants in a single step, nothing more and nothing less. Robots using remote control programs are typically automated, but not robots. It is not possible to classify them as genuine robots.

3. Hybrid programming

Hybrid programming for robots implies that artificial intelligence as well as remote control is integrated within the robotic system. For example, military drones use this type of program. They come with a variety of sensing technology which allow them to search huge areas and spot objects autonomously. But, they require the intervention of a military force to carry out the attack commands against the target.

Humans and Robots: How Do They Interact? The term "human-robot interaction" is a term used to describe the science of investigating how robots communicate with humans. It includes the study of natural language processing that lets robots understand human instructions. In order to even start understanding the process of natural language, we need to be able to comprehend human language that is an longer-standing subject than robotics.

The robotics laws were laid out in the following manner in 1941 by Isaac Asimov, probably one of the best science fiction writers of the past and in 1941:

1. Robots should not cause physical injury to a person. Thus, a robot should not permit a person to suffer harm through its actions.

2. The robot has to be able to obey human orders. It is only possible to not follow instructions if they're violating the laws that was mentioned earlier.

3. Robots must safeguard themselves against total destruction, unless it is in conflict with the two first laws.

Chapter 8: Experiments And Case Studies

1. Dell and Persado

Dell is a global tech company that is famous for its variety of Dell computers. As data plays crucially in Dell's performance, Dell was keen to know what specific words might be key to the success of its marketing emails.

To achieve this goal, Dell joined forces in a partnership with Persado the company which is a specialist in artificial intelligence and machine learning. Persado's AI solution required the creation of efficient marketing content to send out newsletters and emails. Then the AI algorithms focused on data-driven analysis. Every one of Dell's customers received personalised emails.

Their partnership produced outstanding performance. In the beginning Dell's CTR was up by 50. Feedback from customers rose by 46% as did page visits, which rose by 22 percent. In the context of Dell gets thousands of million of page views, the increase of 22% is a significant increase. In addition the 77 % increase in the amount of people who added products to their shopping carts within the category of shopping, which results in an increase in the sales.

Dell extended its AI and machine-learning capabilities to cover the entire marketing and sales ecosystem in the wake of the dramatic rise in customer engagement as well as profits. And Persado looked happy throughout the bank, Dell profited also.

2. Harley-Davidson Motorcycles

The traditional marketing methods were not performing as well as was desired to be for Harley Davidson, a maker of motorbikes that are high-powered. Albert was an AI robot, was introduced by Harley Davidson's New York City branch.

Albert is able to write articles and web pages. Albert can manage marketing campaigns using a wide range of social media as well as emails. Its AI uses machine learning in order to determine the types of customers who are most likely to complete a useful job, like sign up to the mailing list.

Albert is able to achieve this objective by analysing previous customer data and then determining which customers are most likely to stay longer on the site and ultimately, buy.

The bot developed the tests with the segmented approach and realized the possibility of increasing sales by as much as 40 percent. It is able to improve leads by 2,900 percent. Albert was right in estimating that approximately 50 percent of leads will be engaged in the actions.

3. Yelp

Yelp is a business with the goal of sharing a business's and their products real-world reputation. The business gathered a huge amount of pictures from customers when they published review of products and services on the internet for a number of years. The quantity of information collected was impossible to cut through.

Yelp is developing an AI technology that can scan the database of photos in real time by using machine learning. It is able to discern the texture of food items in the pictures looks similar to. It is also able to discern the differences between restaurants that have atmosphere as well as those without ambience by the light in the photo.

So that Yelp's AI has been trained to identify which establishments are suitable for children or have an inviting atmosphere.

4. Vodafone and Uneeq

Vodafone is a major multinational telecommunications corporation with a name that has a huge following among its clients. It has invested a lot of money on AI to ensure that their branding equity did not suffer. It was designed to make consumers see the brand the manner it ought to be perceived.

For building brand loyalty, Vodafone is a firm believer in the quality of customer service. Vodafone's New Zealand arm teamed with Uneeq to design an engaging interactive, welcoming, and hands-on digital persona to accomplish the goal.

The AI is available at a handful of Vodafone outlets across the country. It serves the same purpose like a genuine customer service agent. The representative greets customers with a smiles and helps customers in navigating the prepaid plan they have purchased.

If you purchase, the computer creates the receipt. Vodafone's artificial intelligence is renowned because of its capability to create an authentic conversation. It's nearly identical to what you would expect from a real.

In the end, an AI with the appearance of a human service representative emerges with a voice as well as a face.

5. Trendyol & Liveclicker

Trenyol is an online marketplace. Its sportswear department Trenyol was having difficulties keeping up with international brands like Asos as well as Adidas. It turned to Livecliker which is an AI service.

Liveclicker's software uses machine learning for adjusting marketing messages specific to every customer. The company basically collects the data of users and analyses the data for different preferences. After it's got a general understanding of what each person likes, they create efficient web content specifically designed for the person. It is an excellent approach to make them feel valued.

This method was how Trendyol could create extremely personalized ads.

The different users got various message from different users. Most importantly, purchasers could get their name added to the reverse side of their jersey.

The CTR was increased by 30 percent due to this. Response rates grew by 62 percentage. It's not the worst part, the conversion rate were up by 130 percent.

This is why the business now employs AI and machine learning in order to effectively manage marketing campaigns, targeting various platforms including celebrity endorsements blog posts on SEO and mobile applications in addition to other avenues to interact with its client customers. Its sales have reached an all-time peak!

Chapter 9: The Future Challenges Of Artificial Intelligence

Artificial Intelligence (AI) is one of the key techniques for revolutionizing industries and helping solve global problems. With time increasing technology uses AI as well as Machine Learning algorithms and programs in healthcare facilities, factories and banking, security and various other sectors, and are even extending to e-commerce social media and mobile applications platforms.

But, as AI technology develops and solutions are developed, more issues arise regarding what we can do and whether the resources we have now are sufficient to satisfy changing demands from people. Therefore, in along with the unsolved challenges that remain, the obstacles seem insurmountable and we're still not yet at a point of perfecting our AI systems. These are the most significant challenges to be faced by AI technology.

Processing Unstructured Data

Unstructured data can be extremely beneficial to businesses, but the majority of companies can't obtain significant information from this data because it can't be analysed using conventional methods. It isn't possible to store them within an Relational Database Management System (RDBMS) which is why analysing and processing them is challenging. Examples of data that is unstructured include audio and video files as well as documents, images as well as web-based content.

When, through the application of AI and machine learning non-structured data is decomposed into smaller pieces, then processed and saved as more logical information, it'll make it easier for companies to take informed and rational choices.

The Hunt for AI Talents

In September of 2019, IDC predicted that $97.9 billion would be invested in AI technology by 2023. AI is growing at a an increasing rate, as more and more people are accepting the concept of AI and realize its importance in the digital age.

The rising interest in AI will also result in a rise in the demand for AI technology developers. The evidence is backed by studies of a 74 percent growth in demand for AI experts from 2016 to 2019. But, there's an insufficient supply of AI experts who are able to reach the necessary level of AI deployment and have abilities to design fully functioning systems.

Improving Speech and Text AI

If you operate an online store, then you need to manage a great deal of studying. Feedback from customers, comments, the like, transactions, complaints, reports etc. Through Speech and Text AI, most of the everyday jobs that businesses are required to manage are managed by AI. AI is also able to deliver important reminders or other messages to your contact list. AI can translate text that are in various languages into the your own language. Text and speech AI could be employed by a range of individuals, not just companies.

AI Integration into Cloud

Cloud-based solutions are in great demand. Cloud computing reduces the stress of keeping files and data in your personal device. Instead, it permits you to save them to cloud-based data centers or pools which are also called "clouds". The integration of AI with cloud services dramatically accelerates the operation of the system because it allows it to sort through and handle the massive quantities of data that are stored that are stored in cloud storage. This allows the system to concentrate on other essential jobs.

Privacy

Some mountains or even two of information must be gathered for the purpose of training artificial intelligence to achieve complete machine learning. It is information that comes from real-life individuals. To avoid security breaches, private information must be treated with extreme care, even if the information was not readily accessible.

This issue is as a result of the consent required to use this information. Although the vast majority of individuals have the capacity to decide the extent to which their data will be made available for use in this manner but this may not be the case. If they were aware of what information was being used for, there are some who might not agree to the use of their data.

A few shady companies were in a position to utilize this data at will, and without notifying users of privacy policies.

The number of data breaches that lead to the disclosure of private information is another essential aspect to take into consideration when using or storing sensitive information. In order to protect the privacy of users law enforcement, more strict and punishing legislation is required before data can be used to serve AI to serve purposes. The sale of data should be controlled too, given that it is a legally uncertain field.

Integration to Augmented Intelligence

Nowadays, AI and Data are vital to gain an advantage. They'll likely form as part of a bigger program for process automation and modernization. Using virtual reality in the classroom to instruct learners, a few multinational corporations are experimenting with virtual reality for training simulations as well as boardroom sessions. It's true that this is going to benefit everybody, not just employees. Designers can use augmented intelligence as well as artificial intelligence to develop innovative ways for individuals to collaborate, communicate, or even develop entertainment tools.

Bias

A different issue that is a concern with artificial intelligence concerns bias. Be aware that to identify pattern patterns Artificial Intelligence and Machine Learning require large numbers of data sources. There is a tendency to bias when the data are altered to reflect greater proportions of one population but less of the other.

Consider for instance the ImageNet database for example. It contains a larger percent to white Caucasian faces as opposed to brown and black faces. If an artificial intelligence is educated using this database it can be biased thinking that all faces are white even when there are many.

Different types of data that are representative of all of the population needs to be collected to build the AI. Otherwise, the AI would have the possibility of biases that are built in and requires more intervention than needed in order to rectify.

If an AI algorithm is biased, it may have devastating effects in the future use of the system.

Ownership of Intellectual Rights

If a company creates artificial intelligence AI, the AI system is able to create media that could be considered as digital intellectual property.

While the ability to use digital intellectual property for informational purposes is an excellent quality, some could have negative consequences. A deepfake video produced by an AI system could infringe those rights that were originally granted to the creator. Who is held responsible in the event that the AI software created the video solely on it's own decision?

Chapter 10: The Threat Of To Employment

The effect of AI on job opportunities is an issue that is often discussed and debated. Even though AI could provide new opportunities for employment however, it also has an increased chance of job being lost because of automated processes.

One of the most significant worries is the possibility that AI could totally replace human laborers which could result in massive loss of jobs. The fear isn't unfounded because many fields including manufacturing, transportation, as well as customer service, have experienced significant improvements in technological advancements. As AI technology advances the chances are that other sectors to be able to adopt the same model.

It's not all negative and depressing. AI could also open up employment opportunities for people, specifically in areas such as data analysis, programming or machine learning.

The jobs that require these skills are ones which are sought after and likely to remain even more so in the coming years.

Another possible impact of AI on the job market is the possibility of it leading to a shift in nature of the work humans perform. Since AI replaces mundane and repetitive tasks humans are able to focus on more challenging and innovative jobs that require a human hand. This could lead to satisfaction and satisfaction working experience for many.

The effect of AI on the job market is complicated and complex. Although there's certainly some reason to be cautious but there could be positives to gain. While AI technology evolves in the coming years, it's crucial for both individuals and the entire society to get ready for the shifts which are expected. This means investing into education and training programs to equip individuals with the knowledge they require to succeed in a society in which AI becomes more prevalent.

The Rise of Automation

Automation is the act to replace human labor by computers or programs that are capable of performing similar jobs. Automation is among the major results of the advancement of AI.

The effects of automation are visible in many areas, like manufacturing, logistics and customer service. Thanks to AI the machines and applications can do jobs that were once performed by human beings. The result is an increase in productivity and efficiency these sectors. But the drawback to this is that human beings may soon be becoming obsolete by machines in a variety of jobs and sectors.

Automation's rise has raised issues of inequalities. The most susceptible to losing their jobs due to automation tend to be low-skilled employees within industries like manufacturing, retail, or transportation. If they aren't given opportunities to improve their skills them, they'll be difficult to get a new job.

A more serious issue that is a result of the increase in automation is the risk of AI to take decisions that harm humans. In the case of AI is employed in the justice system, to make decisions concerning sentencing, it might cause unfair and biased decisions. In the same way when AI is employed for healthcare decisions concerning the treatment of patients, it may result in incorrect diagnosis and treatment.

One method to tackle the problems associated with the growing use of automation is by investing in educational and training programs to aid workers in developing the abilities required to collaborate with the machines. Another option is to make sure that AI is created with ethics to ensure that it is unable to adopt decisions that cause harm for humans.

Automation's rise is one of the main urgent concerns in relation the field of artificial intelligence. While AI has the potential to boost efficiency and productivity but it poses serious risk to people working and can cause harmful choices to be automated. It is imperative to take these issues into consideration and develop solutions that will ensure that the advantages of AI can be realized without harming humans.

The Future of Employment

As we've discussed, one of the biggest outcomes of the rapid growth AI is the potential effect on the employment market. A lot of experts have predicted that AI could lead to massive employment displacements, with entire industries becoming automated, and millions of workers being left with no work. It's not an out of the realm scenario as we're experiencing the results of automation in fields such as logistics and manufacturing.

The rise of AI-powered robotics and programs is predicted to be a major impact on the economy. Some experts suggesting that as high as 47% of the jobs will be in danger of becoming eliminated in the coming years. It will be a serious issue for those working in jobs with low skills in addition to those working in fields that are at risk of being automated.

Although it is the case that AI could cause job losses however, it could also provide many new possibilities for employees. Work that requires human interaction as well as problem-solving capabilities are not likely to become automated, so workers who have those skills will remain in demand. Furthermore, job opportunities will emerge within industries that don't exist, since AI continues to transform our lives and work.

In order to prepare for the new world of work in the era of AI employees will have acquire new abilities which are needed in the new digital world. That means they must learn how to operate AI-powered devices and applications and acquiring abilities in fields like the analysis of data, programming and even digital marketing. Businesses will have to fund retraining programs for their employees to ensure they are able to adapt to a changing market for jobs.

The government will also have to adopt measures to combat the negative effects of AI on the job market. It is important to provide the financial assistance needed by workers affected by automation and investing in training and education programmes to assist workers in developing new capabilities.

The future of work in the era of AI is a complicated issue that has both opportunities and risks. It is clear that AI could result in job loss however simultaneously, it can also open up many new possibilities for employees. To be prepared for this new future, both businesses and workers are required to change and acquire new abilities. It is essential for governments to act to help the most vulnerable to the effects of automation.

The Risks of AI in Healthcare

Chapter 11: The Potential For Medical Errors

Potential for medical mistakes using artificial intelligence is an important issue in the field of healthcare. Although AI is able to change the way physicians perform diagnostics and manage patients, it is also the possibility of causing serious injury if used improperly.

One of the greatest risk that comes with AI for healthcare is the potential for mistaken diagnosis or wrong treatment. AI algorithms depend on massive quantities of information for making decisions. However, in the event that data is inaccurate or insufficient or incomplete, the algorithm could come to an untrue conclusion. It could result in serious medical consequences for patients which could include unnecessary procedures or treatments as well as dying.

There is also the risk bias that could present that could be present in AI algorithms. If the data used in order to create the algorithm are or is biased or skewed toward certain groups of people or medical issues it could result in unbalanced outcomes. It could result in disparities between the health outcomes of different patient groups, thereby perpetuating existing disparities, and further aggravating health inequalities.

In addition, there's the possibility that AI devices could fail or fail, causing severe harm to patients. It could happen when the system isn't properly maintained or there are errors or bugs within the program. In certain cases it is possible for the system to get compromised, or even hacked which puts patient data at danger.

In order to address these issues to address these concerns, it is crucial that health professionals as well as AI developers cooperate in order to ensure that AI technology is safe as well as efficient and non-biased. This might require greater test and validation of AI algorithms and also more transparency and accountability the design and development of AI systems.

The potential advantages of AI for healthcare are enormous However, these benefits must be weighed against potential dangers. We all have the responsibility of us to ensure AI can be used with a moral and responsible method, in the best interest of everyone's patients.

The Ethical Implications of AI in Healthcare

As AI is increasingly embedded in healthcare the field, it poses ethical issues that need to be resolved. AI is poised to change the way healthcare is delivered by improving diagnostic, treatment, as well as patient outcomes. However, AI also comes with important ethical issues.

One of the major concern is the risk of AI to continue perpetuating discrimination and bias. If the data used in order to create AI algorithms is biased the outcomes produced by these algorithms will be as well. It will result in health care practices that are discriminatory and harm particular groups of people.

Another ethical consideration is privacy. Because AI is employed to collect and analyse patient information it is possible that the data could be misused or disclosed without consent from the patient. This could result in breaches privacy and confidentiality rights. privacy rights.

There's also concern regarding the effect that AI could have on AI on the doctor-patient interaction. If AI can be used to provide diagnostics or treatment recommendations this could erode the doctor-patient relationship. their patients. Patients might feel they're not being properly listened by or their issues have not been taken seriously.

There is also security concerns about the possibility of AI being used to make money rather than to benefit patients. If AI is employed to take medical decisions based on the financial aspect rather than on the medical need, this could result in less favorable results for patients.

To resolve these ethical concerns in the face of ethical concerns, it's important to set clear guidelines regarding the application of AI for health care. It is important to ensure that the information utilized to create AI algorithms is impartial and secure for privacy of patients, as well as allowing AI to work with a focus on the health of patients over profits.

Furthermore, it's essential to engage patients in the design and use of AI within healthcare. Patients have the right to participate about how their personal information will be used, and they should also be able to give feedback regarding the application of AI for their own care.

As AI advances in healthcare, it's vital to consider the ethical consequences of their use. If we address these issues, we will be able to have AI which is utilized effectively for the benefit of patients while maintaining moral standards for the field of healthcare.

The Dangers of Relying on AI for Diagnosis

One of the major problems with the increased usage of artificial intelligence within medical practice is the use of AI to make diagnoses. Although AI could increase accuracy and speed in the field of medical diagnosis however, there are inherent risks when relying entirely on AI for this critical job.

One of the main risks when you rely on AI to make diagnoses is the possibility of mistaken diagnosis. AI machines can only be as precise in the context of data they've been taught on. If the information they are relying on is flawed or insufficient, the system could produce incorrect results. In addition, AI systems may not be capable of taking into consideration the specific circumstances of every patient, leading to inaccurate diagnoses.

A further risk when using AI to diagnose could be the risk of false positives. AI systems could be too sensitive and detect diseases that aren't in the body, resulting in unneeded testing and treatments. It is not just a waste of precious time and money, but it can even harm patients as they are subjected to unneeded procedures and medicines.

Additionally, depending too much on AI to make diagnoses may result in a decline in the standard of care. Medical professionals as well as other health care providers can become comfortable and depend too much upon the AI system and fail to think about alternative diagnoses or therapies. The result is an absence of customized healthcare and can result in serious consequences for patients.

There is also the possibility of AI systems being compromised or altered. If the AI system becomes affected, it may result in incorrect diagnosis, which could lead in a negative impact on patients. The result could be an erosion of trust with AI devices, which can result in significant consequences regarding how medical diagnoses will be in the near future as well as treatment.

We are aware that AI is able to change the way we diagnose medical conditions, however it is important to understand the risks of relying entirely on AI to perform this crucial mission. It is vital to strike the balance of using AI to reap its benefits and ensuring that medical professionals and doctors are aware and engaged with the particular circumstances that affect every patient. This way, AI can be used appropriately and securely in the medical sector.

The Dangers of AI in National Security

The Potential for AI to be Weapon zed

The possibility that AI could be used for weapons is a scary possibility that everyone should be conscious of. Artificial intelligence is able to adapt and learn in a rapid manner. This implies that it could be programmed to accomplish various activities, which includes ones that could be dangerous or illegal.

Recently, we are witnessing the development of weapons that can be capable of taking decisions by themselves. These weapons are able to create unimaginable destruction and even death. These weapons can be programmed so that they search for and destroy certain targets with no human intervention and leave no space for errors or mercy.

The application of AI to fight cyberwarfare is also an issue. Hackers may employ AI to develop sophisticated cyber attacks that are extremely difficult to identify and fight. Cyberattacks can destroy entire countries, leading to disruption and chaos in a huge way.

AI may also be employed to develop disinformation and propaganda campaigns. Through the analysis of huge amounts of information, AI can create targeted messages that could influence people's opinions and influence elections. This kind of propaganda powered by AI is already being used by a number of nations to influence election outcomes.

One of the biggest concerns that concerns me about AI weaponization, is that it has the risk that it could be misused by the wrong people. State-sponsored terrorist groups and other rogue nations could utilize AI to carry out catastrophic attacks, which could result in immense harm. The rapid growth of AI technology is a sign that AI technology is increasingly available to those who want to employ it for evil motives.

To stop the weaponization of AI We must start now to take action. The government and the organizations need to collaborate to create standards and guidelines that govern the use and development of AI. Also, we must spend money on research and development that will build AI platforms that are transparent responsible, ethical, and accountable.

The threat of AI to be used for weapons is an issue is not something we can afford to ignore. It is imperative to act now to stop the creation and application of AI to harm us. If we don't, it can have devastating implications for the human race.

Chapter 12: The Risks Of Cyber Attacks Using

As AI technology advances and improve, the possibility of cyber-attacks using AI is becoming an issue we need to be aware of. Cyberattacks have always been an issue for both individuals and companies, however thanks to AI the attacks could get more complex and difficult to spot.

One of the main issues with cyberattacks using AI is their capability to imitate human behaviour. Hackers could make use of AI to make convincing fake phishing emails and postings on social networks that seem to be from a reliable source. They can also be specific, by using AI to collect personal data on a victim for making the scam more credible.

A further risk associated with cyberattacks that utilize AI is the capability to make attacks more automated. AI is able to search for networks and system security holes and attack in a way that is automatic. That means attacks can be launched at any time with no need for humans to intervene.

AI may also be employed in the creation of more efficient malware. AI can be utilized to make malware that is able to alter and change, making it more difficult to recognize and eliminate. It is possible for malware to persist to grow and cause harm long after it's been made available.

One of the main worries with cyber attacks that employ AI is the possibility for AI to be able to learn from previous attacks and be more efficient in the future. As AI becomes more prevalent for cyber attacks, these attack will be harder to protect against.

To guard against cyberattacks by using AI both individuals and organizations need to be on guard. It means using passwords that are secure as well as regularly updating software and being alert to fraudulent phishing schemes. Also, it is important to invest in measures to protect your data that recognize and react to AI-powered threats.

Cyberattacks that are able to be carried out made with AI are substantial and should be taken very seriously. As AI grows it is likely that the possibility of cyber-attacks using AI is bound to grow. It is imperative that we be proactive in protecting ourselves and our businesses from these risks.

The Implications of AI Surveillance

The consequences associated with AI surveillance are extensive and extensive that could affect all aspects of our life. As AI technology develops in the field, the capacity to monitor and follow people in a vast scale becomes more feasible. While there could be some advantages for this kind of surveillance such as enhanced security and prevention of crime but there are also serious issues that need to be resolved.

One of the biggest effects associated with AI monitoring is its possibility to abuse control. Companies and governments with the ability to access this tech may employ it to observe and regulate the actions of people, thereby violating the privacy rights and civil liberties of individuals. Utilizing facial recognition technology, as an example it could be used to trace protesters and activists, making it difficult to work without being detected.

In addition, AI surveillance can exacerbate inequality in society. Certain communities, specifically people who are marginalized and under-represented, currently face high amounts of policing and surveillance. The introduction of AI technology can make this issue even more difficult, since algorithms could be biased towards particular groups, which could lead to an increase in discrimination and harassment.

A different area of concern is the possibility of AI surveillance being used for commercial reasons. The technology could be used by companies to gather data about the behavior of individuals and their preferences to target more efficiently the marketing and advertising campaigns. Although this might seem like a harmless practice however, the growing processing of personal information can create important privacy issues as people may not have any control over how their personal information will be used and shared.

The consequences that come from AI surveillance are a complex issue and complex. There are many benefits of the technology, however it's crucial to be aware of the risk and make steps to reduce the risks. It could include the introduction of laws to ensure rights to privacy and civil liberties as well as investing in systems that are open and accountable and participating in discussions with the public concerning the ethical issues associated with AI surveillance. In the absence of these, it can cause significant negative consequences for the individual and to society at large.

The Threat of AI to PrivacyThe Dangers of AI Collecting Personal Information

A further risk associated with artificial intelligence is its capacity to store and collect huge quantities of personal data. Due to the popularity of internet technology and social media, users have become used to sharing their personal details on the internet. Yet, AI takes this to another degree by gathering and analysing the data in a way which humans are unable to understand.

One issue associated with AI taking personal data into account can be the possibility of breach of data. The more data that's obtained and stored, the more valuable for cyber criminals. Cybercriminals can make use of this data for various nefarious motives, such as identification theft, fraud as well as blackmail.

The other concern is the potential of AI being used to perform surveillance. Companies and governments can employ AI to track individuals' actions and tracks their activities. It can be utilized to fulfill legitimate goals including the tracking of criminals. But it is also able to be utilized to pursue more sinister purposes like monitoring opposition politicians.

AI collecting personal information creates privacy concerns. A lot of people are uneasy at the thought of your personal data being taken and used in a way that is not their own. While regulations are that protect people from privacy concerns, the regulations tend to be insufficient and AI is able to find ways to get around them.

Additionally, AI collecting personal information can raise concerns regarding discrimination. AI algorithms can only be in the same way as the information they're trained with. If the datasets utilized to develop AI algorithms are biased that are not reflected in the data, these biases could reflect in the algorithm. This could lead to the discrimination of certain people or certain groups.

The risks of AI using personal information to collect it can't be overlooked. It is imperative that individuals or corporations as well as governments adopt measures to protect the privacy of personal data and ensure that AI is utilized responsibly and in an ethical method. This means implementing strict rules for data protection, assuring it is that AI algorithms are impartial as well as limiting the quantity of personal information kept and accessed. In the absence of this, it could result in serious consequences for people and the society at large.

The Potential for AI to be Used for Targeted Advertising

The possibility of AI to be employed in targeted ads is an issue which has attracted a lot of focus in recent months. The application of AI in the field of advertising is a possibility to transform the field, making it more effective and efficient than it has ever been. But there are important risks with regards to the application of AI in the field of advertising, such as security concerns, the risk of manipulation, as well as the effects on consumer behaviour.

One of the biggest problems with AI for advertising is the possibility of manipulating. In the process of analyzing huge quantities of information about consumer behaviour and habits, AI algorithms can develop extremely targeted advertisements which are crafted to alter consumers' behavior in specific manners. This could be especially problematic in the case of delicate issues like health or politics, as advertisements that are targeted could be utilized to alter public opinion, or even influence behaviour in negative methods.

A further concern with AI in the field of advertising is its effect on privacy. Because AI algorithms gather and analyse massive amounts of data about consumer behaviour, there's the possibility that data may be mishandled or handled incorrectly, resulting in grave privacy issues. This is especially worrying due to recent revelations about security breaches, as well as the mishandling of personal data from companies like Facebook.

However, despite these issues however, there is a significant opportunity to use AI to make effective advertising. When analyzing patterns of behavior and preferences among consumers, AI algorithms can develop extremely targeted campaigns for advertising that are more successful and efficient than traditional marketing techniques. It could result in significant reductions in costs for advertisers in addition to an enhanced and personalized experiences for customers.

To mitigate the risks related to AI to promote advertising, it's essential for regulators and companies to make steps to ensure that consumers are protected from privacy as well as ensure make sure that AI algorithms aren't used or used for a purpose that is harmful. It could mean the development of new rules regarding the application of AI for advertising in addition to adopting standards of best practice and ethics for businesses operating in this area.

The chance for AI to be utilized for targeted ads is substantial and it's essential to consider this new technology cautiously and think about the possible advantages and risks before accepting the concept. There are steps to be taken to ensure that AI will be employed to benefit society without posing security or privacy risks as well as security.

The Risks of AI Being Used for Political Manipulation

Artificial intelligence is a powerful tool to change the face of politics However, it comes with serious risks. Utilizing AI to influence politics is an increasing concern that if left unchecked the issue, it could result in catastrophic implications.

One of the major dangers associated with AI using it to influence politics is propagation of misinformation. AI is able to fabricate credible fake news articles as well as social media posts or even video clips. False narratives could be used to change opinions of the masses, influence elections, and sometimes even provoke violence.

A further risk is using AI to influence the voting process. In the process of analyzing huge quantities of information, AI algorithms can detect potential prospective voters who are swinging and present the voters with targeted messages. They can tailor the messages in order to appeal to the voters' anxieties, preconceptions and needs, thereby influencing the voter to choose the way they prefer.

AI is also a tool to restrict the right to vote. Through analyzing patterns of voting, AI algorithms can identify regions where certain types of people tend to be more likely to vote and then target them with misinformation or other strategies to deter people from voting.

The application of AI to track political activities is a different issue. The government can employ AI to watch messages on social media email messages, as well as other types of communications for signs of potential danger to their system. The monitoring could lead to suppression of freedom of speech and the detention of dissidents from the political establishment.

Chapter 13: The Potential For Accidents And Fatalities

Risk of deaths and accidents can be a major issue when it comes to using artificial intelligence for transportation. As AI grows more sophisticated and is integrated into our everyday life, the chance of fatalities and accidents increases.

The biggest risk could be the usage of automated vehicle. The autonomous vehicles are able significantly reduce the amount of incidents caused by human error. However, they also present new risk. If, for instance, an autonomous vehicle is malfunctioning or commits a mishap and causes an accident that is serious. If hackers could gain the control of these vehicles they could make them into weapons.

It's clear that the possibility of causing deaths and accidents is of great concern in the context of controlling artificial intelligence in the methods of transporting individuals and products. While there is certainly a benefit of AI, it is important to recognize that while there are benefits to AI but it's important to take care and implement measures to minimize the risk. This will require implementing stringent safety rules as well as investing in research and development to gain a better understanding of the risks as well as ensuring that human beings remain in charge of AI machines throughout the day. If we don't take these measures, it could result in a future that sees the positives of AI will be masked by the terrible consequences of accidents or fatalities.

The Ethical Implications of Autonomous Vehicles

The ethical implications associated with autonomous vehicles are profound and must not be ignored. With self-driving cars becoming increasingly commonplace in our streets it is important to think about the ethical issues that arise when using them.

One of the most important ethical questions is that of the liability issue. Who's responsible when an autonomous vehicle gets at fault in an accident? Who is responsible? Is it the maker of the vehicle, the software creator or the person who owns the vehicle? The above questions must be answered before autonomous cars are able to become an everyday mode of transport.

A different ethical concern we've before discussed but is directly related to the effect on work. Due to the rapid growth of self-driving vehicles, many positions in the field of transportation will be replaced, which will leave a lot of workers without a job. It is important to think about the ways we can offer support as well as training to those who are affected by the changes.

Privacy is a further concern. Autonomous vehicles gather a lot of information about the people they are driving with and about their surroundings. These data may be utilized to benefit others, like increasing traffic flow and the prediction of accidents, but it could also be employed for illegal reasons. The data should be secured and used in a responsible manner.

Another ethical problem could be discrimination. Autonomous cars rely on algorithms for making decisions. These algorithms may be biased. As an example, autopilots might be designed to prevent collisions with pedestrians. However, when it is designed to put security of passengers more than pedestrians and make choices that could put pedestrians at risk. It is vital that these algorithms are fair and impartial.

In the final analysis, it is important to take into consideration the effect of autonomous cars on society as overall. Are they likely to increase inequality and enable transportation for all? Autonomous vehicles should be developed in a way that is inclusive of everyone members of the society in mind.

The ethics of autonomous cars as with all issues related to AI are multifaceted and complex. It is important to think about the issues in detail before taking on the technology completely. Autonomous vehicles' benefits should outweigh the risks and should be developed and operated in a moral and ethical manner.

The Dangers of Relying on AI for Transportation Systems

The introduction technology of artificial intelligence (AI) into the transportation system has been described as a major breakthrough. Autonomous vehicles, trucks, and buses will be able to travel on roads and highways with no humans, which reduces accidents and improving effectiveness. But, this advancement in technology is not without risks and should be recognized.

One of the major hazards of relying on AI in transportation is the risk of hacking. Hackers are able to alter the algorithms controlling auto-piloted vehicles, which can cause the vehicles to fail or even crash. It could lead to devastating accidents, the loss of life, or the destruction of property. One incident could be disastrous since it may compromise the entire fleet of vehicles.

A further concern is the reliability of AI when faced with unpredictable circumstances. Although AI can handle huge quantities of data and can make quick rapid decisions, it is not equipped with the agility and flexibility that comes from human drivers. If unexpected hurdles or incidents occur, AI may not be in a position to react properly, which could lead to an accident and other potentially dangerous scenarios.

Also, there is the possibility of reliance too much on AI. As people become more accustomed to vehicles that self-drive and autonomous vehicles, they could begin to believe in the AI technology without question. It could result in complacency as well as a lack of preparation should there be an issue or failure of the system.

The ethical consequences the ethical implications AI within transportation systems should be taken into consideration. Who's accountable if an AI-controlled car causes an accident? Does AI be more concerned with the safety of pedestrians or passengers? pedestrians? This is a complex question that requires an attentive discussion and consideration.

To reduce the risk of these risks It is crucial to include fail-safes in AI technology. regular testing and maintenance is crucial to ensure the system is secure and secure. In addition, ethical and regulatory standards should be developed to ensure that AI is utilized responsibly and ethically.

AI might have the capability to transform the way we travel however, it comes with major risks that have to be taken care of. In recognizing these risks and taking action to minimize the risks, AI can be used fully, and while also minimizing risks to the safety of humans and their lives.

The Threat of AI to Ethics and Morality

The Potential for AI to Make Unethical Decisions

While we work to improve artificial intelligence, it is important to be aware of the possibility of AI to be able to make ethical decisions. AI systems are designed to make choices based on information however, what happens if the data are biased or insufficient? This could lead to AI making choices that can be negative or discriminatory against particular categories of individuals.

A good example of this is the application of face recognition technologies. If the AI algorithm is based on the majority of the data composed of white faces, it could be unable to identify those with darker skin tones. This could lead to incorrect identifications, and possibly wrongful arrests.

A further concern is the usage of AI within the system of criminal justice. AI is used to help predict the chances of someone committing another crime however, these systems have been found to be biased towards certain groups of people. One example is a study that discovered that the AI software employed in the US had twice the likelihood to falsely label the black defendants with greater risk of repeat offenses than white defendants.

The ability of AI to take unethical actions can also be seen in workplaces. Artificial intelligence systems are being employed to aid in the hiring process, however AI systems could bias towards certain types of individuals. As an example the AI system could be programmed to favour candidates who gone to top universities, it isn't an exact indicator of efficiency.

But, what do we do to stop AI from taking unethical choices? One option is to teach AI algorithms on a variety of datasets which accurately reflect the general people. It is also important to be clear regarding the way AI machines make their decisions so that we can spot and rectify errors.

The possibility of AI to take unethical choices is a major concern to be tackled. AI should be programmed to take fair and impartial choices, as well as be open regarding how these decision-making processes are carried out.

Risks associated with AI being used for military purposesOne of the most alarming aspect of the rapidly growing advances in the field of artificial intelligence (AI) is the possibility of its application in military operations. Though some claim that AI might make combat more efficient and safer for humans however, using AI to conduct military activities could result in disastrous outcomes.

It is possible to develop self-contained weapons, also referred to as killer robots. They are devices that recognize and target targets without any human involvement. This type of weapon may lead to a future in which wars are waged entirely through machines, without humans in charge or making decisions.

The application of AI in military operations poses questions regarding accountability and the responsibility. If an AI system makes an error and results in harm then who's to blame? Who are the creators of the AI algorithm? The commanders of the military who employed the system? In the absence of specific answers to these questions can create situations in which no individual is held accountable for the activities of weapons that are autonomous.

A further risk of the use of AI for military purposes is the risk of the system to be attacked or controlled by a group of criminal actors. If a state actor or non-state actor takes the control of an AI-based technology, they can employ it to launch destructive attacks against civilians as well as military targets.

There is also the chance that the usage of AI for military purposes could cause a disruption of the power balance. Nations with high-end AI capabilities may gain considerable advantage over the ones that don't and could result in increased tensions as well as the likelihood of conflicts.

In order to address these issues It is crucial that the policymakers adopt a prudent and responsible approach when it comes to implementation and development of AI for military purposes. Specific guidelines and rules must be implemented to ensure that AI is used to conduct business in a manner that is ethical as well as safe and accountable. In addition, international collaboration is required to prevent the creation of weapons that are autonomous and limit the dangers from AI being employed for military purposes.

It is clear that the usage of AI for military purposes is a worrying technology that has the potential to have devastating effects on global security and stability. It is crucial that decision-makers and everyone else are aware of the potential risks and cooperate in the event that AI will be utilized for purposes that are beneficial to humanity and not put AI in danger.

Chapter 14: The Implications Of The Justice System

The effects of AI for our justice system can be expensive and complicated. AI could change the way we think about legal justice. However, there are serious issues concerning fairness, bias and the privacy of our citizens.

A major implication of AI within our justice system has its possible bias. AI algorithms are the same as the people who created them, and when the information that is used to create these algorithms is biased their results could be biased too. This implies that AI will perpetuate biases that exist in the justice system like racial profiling or discrimination.

Another issue is the lack of transparency within AI decisions. AI systems are typically considered to be black boxes which means it's hard to know how they come to the decisions they make. The lack of transparency in AI systems raises concerns about the accountability of AI systems and due process.

Privacy is a major aspect to consider when considering AI within the legal system. AI systems rely heavily on huge amounts of data which includes personal information for making the right decisions. These data are susceptible to cyber-attacks, including hacking. of cyberattacks. It puts people at danger of having their private data compromised.

However, despite these dangers AI is able to significantly improve the effectiveness and effectiveness in the system of justice. AI tools can aid in identifying patterns of criminal activity as well as assist in the analysis of evidence, and identify the possibility of the occurrence of recidivism.

In order to limit the dangers that could be posed by AI for justice and the justice system, it's essential to consider its use in a cautious manner. This includes conducting thorough testing of AI methods to ensure they're honest and open, as well as using robust privacy safeguards in order to protect personal information.

AI could change the way justice is conducted, however, it raises concerns regarding fairness, bias as well as privacy. It is imperative to approach the use of AI within our justice system with a sense of prudence and take steps to reduce the potential risks while leveraging its power to increase the effectiveness and precision of the criminal justice system.

The Future of AI and Humanity

The Potential for AI to Surpass Human Intelligence

The possibility of AI to outdo human intelligence is an issue which has received a lot of attention recently. Many experts believe that AI could beat human intelligence in the near future, whereas others think that it is extremely likely. One fact is for certain - there is a chance AI will eventually become advanced than human beings.

The notion of superintelligence, an AI that is far superior to humans could be a possibility. It could be achieved via a process called self-improvement that recurses, where an AI enhances its own intelligence, and grows cleverer each time it is upgraded. In the end, it may soon reach the level of sophistication superior to that of humans.

The implications of superintelligence are immense. Superintelligent AIs could take the right decisions and make actions that go beyond human understanding, and can have significant consequences for the human race. As an example, a superintelligent AI might decide humans pose the biggest danger to its existence, and then take actions to eradicate humans.

There's also a risk about the possibility that superintelligent AIs create goals that aren't in line with human values. In particular, it may place safety over efficiency and cause disastrous outcomes. It is referred to as the "alignment problem" and is one of the major issues confronting AI the development of AI.

It is crucial to remember that we're still not in the process of developing an artificial intelligence that can be superintelligent. It is nevertheless crucial to think about the potential implications of this technology, and devise strategies to reduce possible risks. This includes developing AI technology that is transparent and accountable. This will ensure humans are central to AI creation.

The possibility of AI to outdo human intelligence is something which cannot be ignored. Although the creation of a superintelligent AI isn't yet an option, however it's important to be aware of the implications and formulate strategies to mitigate these threats. While we create AI technology, it's essential to ensure that human safety is the top priority and ensure that the advantages of AI do not exceed the risks that could be posed by AI.

The Ethical Considerations of Creating Superhuman AI

As we advance the development of Artificial Intelligence, we could be soon developing superhuman AI, machines capable of beating human beings in almost every job. Before we dive forward into this exciting future there are ethical concerns that we must be aware of.

One of the major problems with superhuman AI is the possibility for it to be in control. When it is created, the superhuman AI may soon grow so sophisticated that human cannot comprehend or manage it. It could have disastrous effects, since AIs could choose to behave in ways which are detrimental to humankind.

Another factor to consider is the impact the superhuman AI might have on the employment market. If AI is competent enough to do every job better and more effectively than human beings, it will result in widespread employment and disruption in the economy. This could have wide-ranging implications for both the political and social spheres.

One of the biggest ethical issue in the creation of artificial intelligence that is superhuman in the context about whether it's morally acceptable to develop machines smarter than human beings. Many argue that it would violate the human rights of humans, whereas others claim that it's the inevitable course of technological development.

The ethical issues for creating AI that is superhuman will be debated in the years ahead. While we work to create artificial intelligence, it is important to be aware of the possible negatives and the benefits of each advancement.

The Implications of AI on the Future of Humanity

The ramifications of AI for the future of mankind are enormous and in that sense, very complicated. AI has the capacity to change the course of human evolution. In the meantime, as AI technology develops in a rapid pace and speed, it's becoming obvious that we are moving into new territory. Although the advantages associated with AI have numerous benefits and are unquestionable however, the dangers and threats are significant as well.

The most important issue could be the loss of jobs as well as the automation of several industries. When AI systems get advanced and more capable in the future, they'll take over human laborers in numerous areas, which will lead to huge employment and social disruption. The implications of this will be profound for society, the economy and how people live their lives.

A further impact is the chance to allow AI to be super intelligent and even surpass human intelligence this could lead to formation of an rogue AI that could pose a grave risk to mankind. This is often referred to by the name of "AI singularity," and it's a subject that is the subject of heated debate between AI scientists and researchers. Many believe it's unavoidable, while other argue that it's unlikely, or perhaps impossible.

There are other concerns, including the possibility to allow AI to be utilized to commit crimes for example, cyber-attacks, or even the development of self-contained weapons. It could result in a whole new world of conflict and creating an enormous threat to world security. There is also the possibility of AI using it to further create social inequality and biases that lead to being further excluded from already vulnerable populations.

However however, there are numerous potential advantages to AI including enhanced healthcare, better efficiency of transport, and improved research. But it's crucial to approach AI cautiously and think about the threats and risks simultaneously. It is essential to insist in the simplest terms, that AI is created with a moral and responsible method, with a particular focus on maximising its potential benefits and keeping the potential for harm to a minimum.

The consequences of AI regarding humanity's future are enormous, and it is crucial to be aware of all the implications, both positive and negative, carefully. AI is poised to transform many areas of our daily lives. There are also significant risk and potential dangers that need to be addressed. In the new age of technological advancement be sure to be cautious and put security and the well-being of the human race above all other considerations. It could be a mistake in the event that we do not.

Chapter 15: The Early Beginnings Of Artificial Intelligence

"I think AI will change the world more than anything in the history of mankind."

Kai-Fu Lee

Humans have been creating and innovating instruments since the beginning of the age. From the creation of the wheel until the evolution of language the first humans utilized both their surroundings as well as their imagination in order to solve everyday issues. From primitive hunting equipment to the building of shelters Our ancestors displayed the ability to invent and invention. Additionally that many of the innovations have utilized a type of artificial intelligence, for instance for when the first humans utilized the fire to cook food, and sharpened rocks to create tools. Early use of AI enabled our forefathers to satisfy their requirements, which ultimately shaped our lives in the present. Utilizing AI to make use of the power of the natural environment early human beings set the foundation for the creation of artificial intelligence later decades.

Technology's evolution has played an important role in the development in the field of Artificial Intelligence (AI). Since the beginning of human civilization, people have looked for ways to make their lives more convenient and efficient. It was often about finding innovative methods of using the tools and technology that were already available like the development of the automobile. As societies developed and expand, so did the tools used in order to reach their goals. This led to the development of innovative technologies and innovations including the steam engine which set the stage for industry's revolution. The advent of new technologies like the radio and telephone significantly influenced the ways people communicated that influenced the growth of AI. With the advent of new technology and new methods of thinking were developed, so did the ways to think, allowing for the creation of advanced AI algorithms. At the turn of the 20th century technological advancements were at the point that the first computers were developed and the very first AI software was developed.

Artificial intelligence (AI) is an ongoing event throughout the course of human history. AI dates through the ages in the quest for instruments that could automate and improve the processes. At the beginning it was a matter of instruments like the abacus that was utilized to aid in automating basic computations. With the advancement of technology as did AI. The 19th century was the time when the science of AI has expanded to encompass the study of machines which can be programmed and behave as humans. This was possible due to the invention of computers and programming languages that allowed computers to manage and store information. AI technology like machine learning as well as natural language processing were also beginning to appear in the latter part of the decade which allowed computers to gain knowledge from the data they collect and to interact with human beings in meaningful ways. AI is now omnipresent throughout our modern world, and has vast implications on how people live, work and even interact.

AI is present in a variety of shape or form since the beginning of time according to many artifacts, documents as well as other evidence from the past. For instance, in the early days of Greece, Aristotle wrote about developing machines that would function and think like humans. Similar to that, during the 4th century BC in China, The Chinese philosopher Mozi utilized the concept of "yin-yang" to develop a machine that could detect patterns and then make choices. Although they were not modern-day technology could perform easy tasks, like processing and sorting information.

At the time of the Middle Ages, the Islamic world was able to make significant advancements on AI technology. Mathematicians like Al-Khwarizmi created algorithms to perform calculations. These were the precursors to the modern AI. The 12th century was when mathematician Fibonacci invented the Fibonacci sequence that remains in use to create AI in the present. In the meantime, the Renaissance saw the creation of mechanical automata, also known as devices that were able to operate independently of human intervention. While they're primitive, could accomplish basic tasks for example, opening doors and playing instruments.

The 20th century was when AI took on more advanced forms. It was in 1950 that British mathematician Alan Turing came up with the famed "Turing Test", which was created to test an AI's capacity to reason and think like humans. This test established the basis of modern AI which was the catalyst for the development of advanced algorithms and computer software. In the present, AI has become a regular part of our life such as facial recognition and autonomous cars.

The development of AI is a continuous process since the beginning of time however its influence on the world has risen exponentially in the last few years. AI has changed how we interact with machines, and it has allowed us to achieve breakthroughs in fields like transport, healthcare, and entertainment. AI is also having significant influence on our economy and helped us develop new industries and jobs. While AI grows in the future, it's clear its influence on our society will continue to expand.

The effect that AI (AI) on ancient societies was vast and profound. Although AI technology was in its early stages and was being utilized with incredible effect by early civilisations across a variety of areas. As an example, AI was used to create sophisticated algorithms that assisted in the forecasting of weather. It allowed the ancient people to prepare and plan for the coming years. AI could also be used to create complex systems of farming, which allowed greater efficiency in distribution and production of food. Additionally, AI was used to examine large volumes of information and take choices, which allowed ancient civilizations to better make utilization of their resources.

The effect of AI in prehistoric societies did not stop at practical uses. AI could also be used in the development of intricate religious and philosophical systems that provided a basis that allowed for understanding the world as well as those around it. AI was also utilized to build efficient tools for communicating that allowed for the dissemination of ideas and knowledge over vast distances. Overall, AI played an essential contribution to the growth of the prehistoric civilizations, and its effects continue to be felt even today.

Chapter 16: Exploring The Impact

"AI is not a science fiction thing, it is already changing our lives."

Fei-Fei Li

The Renaissance was a time with a great amount of discovery and change from the 14th through 17th century. In this period, Europe saw an explosion of knowledge and creativity as well as advances in art, science as well as philosophy, literature as well as technology. It was the Renaissance was a time that saw exploration and rediscovery and the world was became open to new concepts and views. The Renaissance saw the development of some of the most influential thinkers such as such thinkers as Galileo, Descartes, and da Vinci, who shaped our thinking and the way we do life to this day. AI played a part in this time when technological advances allowed to create machines that were able to think as well as analyze and make the right decisions. AI helped to broaden knowledge as well as to streamline complex processes which allowed humans to concentrate on a more imaginative and worthwhile activities. The Renaissance revolutionized the course of history and AI played a major role in the development of the revolution.

The Renaissance period was a time with a lot of progress and change in the realms of sciences, art, as well as technology. AI was just one of several technological breakthroughs that occurred during this period. The growth of AI in the Renaissance era was mostly motivated by the necessity to streamline tedious and lengthy jobs. An example of this is the creation of algorithms that could be used to automatize mathematical computations. Mathematicians could focus on more complicated questions. It was evident as the creation of automated looms which made it possible to mass-produce of textiles. AI was also employed to develop complex calculations for astronomy as well as develop innovative techniques for exploration and navigation. The development of AI at this time laid the basis for further advancements in this area and made a an impact that was significant in technology and the Scientific Revolution.

AI was first able to exert an impact in the revolution in science that took place during the seventeenth century. In this period advancements in technology and discoveries in science were inspired by advancements in AI. One of the greatest examples is the creation of automated machines that were enabled by the creation algorithmic algorithms based on AI. The algorithms helped in the creation of brand new and more efficient machines which then enabled growth of many scientific disciplines. In the case of Isaac Newton, for instance, his creation of calculus was made possible through the creation of an AI-based algorithm which could compute complicated equations. AI has also helped in the creation of novel theories and research including the research that was done by Galileo Galilei and Johannes Kepler using AI-derived calculations to make their breakthroughs. AI was also able scientists to predict accurately the movements of planets. This feat could not have been achieved without AI. Furthermore, AI allowed for the creation of treatments for medical conditions as well as other medical technology including the creation of microscopes. The advancements in AI had profound impact on the development of science and helped shape the way we view as it is today.

The Enlightenment period that began at the end of the 17th century, and continued until the end of the 18th century was a time of massive change in values and ideas that significantly influenced the ongoing advancement of AI. One of the leading philosophers of the period included Rene Descartes, who developed the notion of an "intelligent machine in his work The

Discourse on the Method. Descartes believed that machines were able to be programmed to perform actions and think in a rational way as well as he believed machines were able to be programmed with the characteristics of human ability. This notion of machines that were intelligent was further refined by Gottfried Leibniz who was the first person to utilize the word 'calculus'. Leibniz also invented an early model that was a type of mechanical calculator that laid the basis for the earliest computers.

The Enlightenment period also marked the time of tremendous advancement in math and physics. This provided the basis for the creation of AI. Isaac Newton developed the laws of gravity and motion that allowed the creation of algorithms to compute complicated equations. Additionally, mathematician Leonhard Euler devised the first algorithm to solve linear equations. It is a major revolution in the field of AI.

The Enlightenment period marked a time with a lot of scientific and intellectual growth, and advancements in the fields of mathematics, physics and computer science established the foundations for the growth of AI. The concepts that were developed by Descartes, Leibniz, Newton and Euler helped shape the evolution of AI and opened the door to future generations of engineers and scientists to build ever more complex and efficient machines and algorithms. The influence from the Enlightenment period in the creation of AI can't be overstated and the legacy of its influence lives on with the current AI tools and techniques we utilize in the present.

The influence on the impact of AI in the Renaissance was widespread and profound. From new and innovative technologies to the philosophical implications AI was a major influence across all aspects of our lives in the period. Automating routine tasks, AI allowed scientists and philosophers to concentrate their attention and energy on more difficult questions. AI has also triggered technological and social change because new technology enabled individuals to improve their efficiency as well as quality of their lives. AI was also a major impact on the philosophy of mind, since the idea of man-made intelligence challenged the traditional views concerning the nature of information and learning. AI was a significant influence on the growth of the scientific method that relied on experiments and analysis to unravel the principles of nature. In addition, AI had a profound impact on the arts since musicians and artists started exploring new methods to express themselves and create. At the time in the Renaissance, AI had established itself as an influential influence throughout Europe in shaping and altering the history of Europe throughout the centuries that followed.

Chapter 17: The Industrial Age

"AI is going to change the world. We must get started on it.

Sebastian Thrun

The Industrial Revolution of the late 18th and 19th centuries was a pivotal moment in the development of production, technology and industrialization, and served as the catalyst that led to an unparalleled period of economic growth as well as advancement. The period witnessed the development and use of new machines as well as the usage of electric power for power and the emergence methods of mass production. Due to these advancements, manufacturing of goods grew exponentially which led to a greater need for workers. This increased demand for labor triggered the shift away from manual work to an efficient industrial process, which eventually led to the rise of artificial intelligence, and its effect upon our lives in the Industrial Age. In this time, we also witnessed the creation of modern communications systems as well as transportation networks that enabled the swift transfer of services, goods, as well as information across the globe. This was the Industrial Revolution, then, was a pivotal moment in the evolution of technological progress and the implications it had for society.

The Industrial Revolution of the late 18th century and the early 19th century witnessed a period of unparalleled development, growth and innovation throughout Europe as well as North America. The world was experiencing advances in science, technology, as well as industry, it saw the advent the concept of artificial intelligence (AI). In this time, researchers and innovators started to look at the possibility of creating machines which could learn and think independently. While only in its beginnings was the basis to develop AI over the decades to follow. One of the earliest examples of AI was the

Charles Babbage's Difference Engine, a mechanical calculator which could tackle mathematics without the intervention of humans. In the course of time, as the Industrial Revolution progressed, the progress of AI was accelerating, with scientists looking for new ways of automating processes and developing machines that mimic human behaviors. In the late 18th century AI became an important component of industrialization and will continue to make a profound effects on our society for the foreseeable future.

At the time of Industrial Revolution, the development of technology as well as the need for labour increased dramatically. In this period, we witnessed the dawn of the industrial revolution as well as the rise of automation as well as artificial intelligence (AI). When the industry grew as was the demand to have machines capable of performing difficult tasks in a hurry and effectively. AI was the ideal solution to this challenge that allowed machines to assume the duties of human workers as well as allowing them to operate on their own and independently.

One of the major advancements of this time was the creation of a computer that could be programmed and soon it became a vital component of the manufacturing process. In the process of storing information and directions within a computer system the machines were able to be programed to perform particular tasks, like the sorting of items as well as controlling machinery and formulating complicated mathematical equations. This led to better production efficiency as well as greater precision in production.

Additionally, AI also enabled machines to assume more complicated functions, including surgical diagnosis or robotic surgery. Through the use of AI computers could learn to identify patterns and react to certain circumstances. These allowed them to carry out higher-level tasks than previously and provided opportunities for new applications of AI across a range of fields.

All in all, AI and automation had significant impact on industrial age. Through replacing manual labor through machines, it made to speed up and improve manufacturing procedures. However it allowed machines to assume more complicated functions, which ushered in the new age of technological development and pave the way for further development of AI.

The influence on the impact of AI upon the Industrial Age is undeniable and is a key component of the evolution of industrialized economies in the present. The 18th century saw steam-powered devices first appeared on the scene. It began the Industrial Revolution, a period that saw rapid technological advancements and economic expansion. When the Industrial

Revolution progressed, so as the advancement of AI technology, which was utilized to boost efficiency and productivity in manufacturing processes. Artificial intelligence-driven automation was able to replace manual labor across a variety of industries including manufacturing, transportation and even food production which allowed for greater efficiency in production as well as delivery of products and services.

Alongside improving the productivity of workers, AI also had a major influence on the structure of the labor force. Artificial intelligence-driven technology resulted in job loss in a wide range of fields due to the substitution of human labor by machines. The change in the structure of work led to a whole new category of employees that had to compete against AI-driven automation to get positions. This in turn affected the wage rates, since companies could pay lower in wages because of the improved productivity of artificial intelligence-driven automation.

AI has also had an important effect on the economy in general. Lower wages that resulted from the automation of AI allowed more purchasing power for people, which resulted in a higher demands for both items and services. This ultimately, triggered an increase in economic growth and the growth. AI helped companies develop new services and products and stimulated the economy.

In the end, AI has had a major influence upon our lives in the Industrial Age, transforming the working environment and economic activity in general. Automation driven by AI has led to an increase in efficiency. This has led to cheaper prices and more purchasing capacity for the consumer. Additionally, it has resulted in loss of jobs, since many work-related jobs are substituted by automated processes. However, despite these negative effects, AI has also facilitated expansion and growth in the economy, developing new services and products which stimulate growth in the economy as well as development.

The emergence of AI into the modern age was a major effect on the economic. Before the advent of AI it was determined largely through manual work, with humans performing the majority of jobs. With the rise of AI it was evident that many of the tasks which were performed by human beings were now possible to automatize and thereby allowing businesses to lower the cost of labor and boost efficiency. It had a wide-ranging impact, since it enabled companies to boost their earnings while decreasing their expenses for labor. AI allowed companies to create goods and services efficient and quickly, allowing their businesses to have an advantage in competition with their counterparts.

One of the major impact of AI for the economy was the change towards automated processes. It allowed businesses to cut down on their costs for labor as they were no longer required to pay for workers' manual work. It also allowed companies to make more money, since they can now create products and services faster and more efficiently. Furthermore, the advent of AI resulted in the creation of new businesses including robots, computer science and AI. These new industries provided work opportunities and jobs for the workforce which led to a more prospering economic system.

Chapter 18: Modern Society

"The pace of progress in artificial intelligence is incredibly fast."

Elon Musk

The 20th century marked an era of unimaginable changes, with advancements in technology being made at an unimaginable speed. The 20th century witnessed the advent of supercomputers as well as the growth in the Internet, and the advent AI (AI) and the beginning of the age of digital. The 20th century saw the globe be increasingly connected thanks to the growth of global communication technology and the spreading of revolutionary ideas and new technology. The world witnessed a significant rise in the number of city dwellers along with the growing of the world economy. Additionally, the technological advances of the time had an enormous effect on how we reside, work and communicate with one another. From the advent of automobiles through the advent of internet technology, the developments were a major influence on our society, changing how we interact with their surroundings.

The 20th century was marked by an abrupt shift in the growth in the field of artificial intelligence (AI) beginning at its inception around 1940, and the current widespread acceptance in society. The domain of AI was able to experience a period with rapid growth that began in the 1950s and lasting to the late 1990s. The 1950s were the time when AI researchers began exploring the possibilities of computers being used to complete difficult task. The 1950s saw the introduction of numerous crucial AI software, including ELIZA and Stanford Research Institute's General Problem Solver (GPS). The 1960s were when AI researchers were able to apply their techniques to a wider variety of projects, such as the field of robotics, natural-language processing, as well as computer vision. In this time, we saw the rise of a number of the first expert systems. These were machines which could mimic the knowledge that a person would have in various fields. The 1970s saw AI researchers focused their efforts on creating advanced methods for machine learning, including neural networks as well as decision tree models. The 1980s saw AI was a significant field of research, with researchers investigating the possibilities of AI across a broad range of fields, including medical diagnostics and autonomous automobiles. In the 1990s, we saw the development of high-performance AI methods, such as deep learning. This enabled machines to be taught from huge data sets, and allowed the machines to complete difficult tasks with unimaginable precision.

The technological revolution that took place in the mid-20th century, that led to the development of computers for personal use, internet, as well as a myriad of other technological advancements and innovations, had an enormous impact on the growth in artificial intelligence (AI). As new technology emerged and grew, so did the new possibilities for AI to grow and become stronger. The ability of AI to automatize procedures, enhance decision-making as well as enhance the capabilities of humans became apparent quickly. Thanks to the development of advanced computers, scientists could develop AI algorithms that were able to analyse and process data more quickly than before. AI-driven technology, like deep and machine learning, helped computers "learn" from data and take decisions with no human input. As AI capabilities increased as the impact it had on society. AI-powered technology began being utilized in various fields, ranging from finance to healthcare, as well as in homes. AI is present all over our lives, from autonomous vehicles to virtual assistants and intelligent home appliances. As AI is advancing and becoming more sophisticated in the future, it is bound to be more influential on our lives as well as the ways we interact and utilize technology.

The effect of AI in our modern world has been substantial. From the invention of computers as well as the creation of the internet through the automation of manual tasks as well as the creation artificial intelligence, the speedy advancements in technology have profoundly impacted on the way we live. AI allows us to achieve more tasks with much less effort and time and also has given us an unprecedented level of accessibility to data and information. This has allowed us to simplify many processes as well as increase efficiency and lower expenses.

AI is also revolutionizing how we communicate and communicate with one another. Artificial Intelligence is currently being utilized to develop virtual assistants like Alexa as well as Google Home which are able to respond to questions, offer data, or even operate gadgets. AI can also be employed to improve sales and customer service in order to enhance customer service and increase engagement. Chatbots powered by AI are used to streamline conversations, and also provide responses to questions from customers.

AI has also had an impact on the ways we do business. Automated and AI-based technology is utilized to simplify routine tasks, reduce the time spent to free up workers from their busy schedules to focus on more productive job. Businesses employ AI-powered software to take data-driven decision making to analyze customer behaviour, and improve processes.